ACTIVE CARBON

ELLIS HORWOOD SERIES IN PHYSICAL CHEMISTRY

Series Editor: Professor T. J. KEMP, Department of Chemistry, University of Warwick

Atherton, N.	Electron Spin Resonance Spectroscopy
Ball, M. C. & Strachan, A. N.	Chemistry and Reactivity of Solids
Cullis, C. F. & Hirschler, M.	Combustion and Air Pollution: Chemistry and Toxicology
Davies, P. B. & Russell, D. K.	Laser Magnetic Resonance
Horvath, A. L.	Handbook of Aqueous Electrolyte Solutions
Jankowska, H., Świątkowski, A. & Choma, J.	Active Carbon
Jaycock, M. J. & Parfitt, G. D.	Chemistry of Interfaces
Jaycock, M. J. & Parfitt, G. D.	Chemistry of Colloids
Keller, C.	Radiochemistry
Ladd, M. F. C.	Symmetry in Molecules and Crystals
Mason, T. J. & Lorimer, P.	Sonochemistry: Theory, Applications and Uses of Ultrasound in Chemistry
Navratil, O. *et al.*	Nuclear Chemistry
Paryjczak, T.	Gas Chromatography in Adsorption and Catalysis
Quinchon, J. & Tranchant, J.	Nitrocelluloses
Sadlej, J.	Semi-empirical Methods in Quantum Chemistry
Snatzke, G., Ryback, G. & Slopes, P. M.	Optical Rotary Dispersion and Circular Dichroism
Southampton Electrochemistry Group	Instrumental Methods in Electrochemistry
Wan, J. K. S. & Depew, M. C.	Polarization and Magnetic Effects in Chemistry

BOOKS OF RELATED INTEREST

Balzani, V. & Scandola, F.	Supramolecular Photochemistry
Buxton, G. V. & Salmon, G. A.	Pulse Radiolysis and its Applications in Chemistry
Epton, R.	Chromatography of Synthetic and Biological Polymers: Vols 1 & 2
Harriman, A.	Inorganic Photochemistry
Hearle, J.	Polymers and their Properties: Vol. 1
Horspool, W. & Armesto, D.	Organic Photochemistry
Kennedy, J. F. *et al.*	Cellulose and its Derivatives
Kennedy, J. F. *et al.*	Wood and Cellulosics
Krestov, G. A.	Thermodynamics of Solvation: Solution and Dissolution; Ions and Solvents; Structure and Energetics
Lazár, M. *et al.*	Chemical Reactions of Natural and Synthetic Polymers
Milinchuk, V. K. & Tupikov, V. I.	Organic Radiation Chemistry Handbook
Nevell, T. J.	Cellulose Chemistry and its Applications
Šičpek, J. *et al.*	Polymers as Materials for Packaging
Švec, P. *et al.*	Styrene-based Plastics and their Modifications

ACTIVE CARBON

HELENA JANKOWSKA Ph.D., D.Sc.
ANDRZEJ ŚWIĄTKOWSKI Ph.D., D.Sc.
JERZY CHOMA Ph.D., D.Sc.
all of the Military Technical Academy, Warsaw, Poland

Translation Editor:
T. J. KEMP
Department of Chemistry, University of Warwick

ELLIS HORWOOD
NEW YORK LONDON TORONTO SYDNEY TOKYO SINGAPORE

English Edition first published in 1991
in coedition between
ELLIS HORWOOD LIMITED
Market Cross House, Cooper Street,
Chichester, West Sussex, PO19 1EB, England

A division of
Simon & Schuster International Group
A Paramount Communications Company

and

WYDAWNICTWA NAUKOWO-TECHNICZNE
Warsaw, Poland

© 1991

Translation: © Ellis Horwood Limited

All rights reserved. No part of this publication may be reproduced, stored in a retrieval system, or transmitted, in any form, or by any means, electronic, mechanical, photocopying, recording or otherwise, without the prior permission, in writing, from the publisher.

Printed in Poland

British Library Cataloguing in Publication Data

Jankowska, Helena
Active Carbon
1. Carbon
I. Title II. Świątkowski, Andrzej III. Choma, Jerzy
546.6811
ISBN 0-13-004912-3

Library of Congress Cataloging-in-Publication Data available.

Table of contents

Chapter 1 Introduction 9

Chapter 2 Production of Active Carbons 13
  2.1 General 13
  2.2 Raw Materials for the Production of Active Carbons 15
    2.2.1 Hard Coal 19
    2.2.2 Raw Materials with a Low Degree of Carbonization 23
    2.2.3 Binding Materials for Preparing Extruded Active Carbons . 28
  2.3 Technological Operations in the Production of Active Carbons . 29
    2.3.1 Granulation 29
    2.3.2 Carbonization 31
    2.3.3 Activation 38
  2.4 Estimation of the Properties of Active Carbons 52
    2.4.1 Collection and Preparation of Samples 52
    2.4.2 Physical Tests 53
    2.4.3 Adsorption Tests 56
    2.4.4 Chemical and Physico-Chemical Tests 60
  2.5 Characteristics and Applications of Commercial Active Carbons . 61
  References 71

Chapter 3 Structure and Chemical Nature of the Surface of Active Carbon . . . 75
  3.1 Molecular, Crystalline and Porous Structure of Active Carbon . 75
  3.2 Non-Carbonaceous Additives Present in Active Carbon 80
  3.3 Chemical Nature of the Surface of Active Carbon 81
    3.3.1 General Remarks 81
    3.3.2 Surface Functional Groups: Nature and Types 82
    3.3.3 Methods of Analysis of Surface Functional Groups Containing Oxygen 85
  3.4 Effect of the Chemical Nature of the Active Carbon Surface on its Adsorption Properties 96
    3.4.1 General Remarks 96
    3.4.2 Adsorption of Polar Adsorbates from the Gas Phase 96
    3.4.3 Adsorption of Non-Electrolytes from Binary Solutions . . . 99
    3.4.4 Adsorption of Electrolytes, and Electrode Properties of Active Carbons 100
  References 104

Table of contents

Chapter 4 Models of Adsorption and their Corresponding Isotherms 107
   4.1 General Description of the Phenomena 107
   4.2 Henry's Law 108
   4.3 Langmuir's Isotherm 110
   4.4 The Brunauer, Emmett and Teller (BET) Isotherm for Multilayer Adsorption 113
   4.5 The Harkins–Jura Relative Method 120
   4.6 Capillary Condensation of Vapours and Evaporation of the Condensate 121
   4.7 Potential Theory of Adsorption 128
   4.8 Adsorption on Active Carbons with Micro-, Meso-, and Macropores . . 132
References 138

Chapter 5 The Theory of Volume Filling of Micropores and its Developments . . 141
   5.1 Foundations of the Theory 141
   5.2 Developments of the Micropore Filling Theory 144
   5.3 Heterogeneity of the Microporous Structure 145
   5.4 Filling of Heterogeneous Microporous Structures 146
   5.5 Surface Area of the Micropores 153
   5.6 Thermodynamics of Adsorption in Micropores 155
   5.7 Complete Analysis of the Porous Structure of Adsorbents 158
References 161

Chapter 6 Energy Effects in Adsorption 163
   6.1 Heat of Adsorption 163
   6.2 Some Types of Calorimeters 167
References 168

Chapter 7 Adsorption from the Liquid Phase 170
   7.1 Fundamental Relationships for the Surface Layer 170
   7.2 Gibbs Adsorption Isotherm 172
   7.3 Isotherm for Adsorption by a Solid from Binary Liquid Solutions . . 175
   7.4 Adsorption on Active Carbons 180
References 190

Chapter 8 Techniques for Testing the Porous Structure of Active Carbons . . 193
   8.1 Adsorption Methods 193
      8.1.1 General Remarks 193
      8.1.2 Sorption Spiral Balances 194
      8.1.3 Adsorption Liquid Microburettes 196
      8.1.4 Adsorption Manostat 200
      8.1.5 Automation of Adsorption Testing Apparatus 205
      8.1.6 Methods of Determining Adsorption under Dynamic Conditions . . 206
   8.2 Total Pore Volume. Real and Apparent Density 207
   8.3 Mercury Porosimetry 212
   8.4 X-Ray Small-Angle Scattering Method 214
   8.5 Electron Microscopy 217
References 218

Chapter 9 Applications of Active Carbon 219
   9.1 Adsorption from the Gas Phase 219
      9.1.1 General Remarks 219

Table of contents

- 9.1.2 Purification of Industrial Exhaust Gases, Recovery of Valuable Components, Separation of Gas Mixtures 219
- 9.1.3 Applications of Impregnated Active Carbons for Protection of the Upper Respiratory Tract against Toxic Substances 225
- 9.2 Adsorption from the Liquid Phase 231
  - 9.2.1 Food Industry 231
  - 9.2.2 Water and Wastewater Treatment 237
  - 9.2.3 Chemical and Pharmaceutical Industries 245
  - 9.2.4 Medicine 246
- 9.3 Active Carbon as an Electrode Material 250
- 9.4. Utilization of Ion-Exchange Properties of Active Carbon 254
- References 257

Chapter 10 Regeneration of Spent Active Carbon 260
- 10.1 Theoretical Basis of the Regeneration and Reactivation Process . . . 260
- 10.2 Conditions of the Regeneration and Reactivation Processes 263
- 10.3 Methods of Regeneration 265
  - 10.3.1 General Comments 265
  - 10.3.2 Thermal and Gas Methods 265
  - 10.3.3 Extraction and Chemical Methods 268
  - 10.3.4 Electric and Electrochemical Methods 272
  - 10.3.5 Other Methods of Regeneration 273
- 10.4 Regeneration of Impregnated Carbon Adsorbents 274
- References 276
- **Index** 279

CHAPTER 1
Introduction

There are as many reasons to call carbon a 'conventional' material as there are to consider it an 'advanced' one. Over the last three decades industry has played a leading role in the utilization of 'conventional' carbon materials in the form of arc furnace electrodes and various other electrodes for electrochemical processes. The 'advanced' carbon materials such as synthetic diamonds and carbon composites obtained from carbon fibres are still finding their way into wide practical use. Among the known carbon materials, the role of active carbon cannot be overestimated, either in modern industrial practice or in everyday life. The physical and chemical properties of active carbon are so fascinating that many researchers have become captivated by this material.

Active carbon is a processed carbon material with a highly developed porous structure and a large internal specific surface area. It consists, of course, principally of carbon (87 to 97%) but also contains such elements as hydrogen, oxygen, sulphur and nitrogen, as well as various compounds either originating from the raw material used in its production or generated during its manufacture. Active carbon may also contain various useless mineral substances in quantities of 1 to 20%. These substances are often removed, when the content of the so-called ash decreases to 0.1 to 0.2%. Active carbon has the ability to absorb various substances both from the gas and liquid phases. It is this ability to arrest different molecules at the inner surface of active carbon that justifies calling it an adsorbent, and a very powerful adsorbent indeed it is. The pore volume of active carbon usually exceeds 0.2 $cm^3\ g^{-1}$ but in many instances it is greater than 1 $cm^3\ g^{-1}$. The inner specific surface area is generally greater than 400 $m^2\ g^{-1}$ but often exceeds this value, reaching 1000 $m^2\ g^{-1}$. The linear dimensions of the pores (i.e. their radii) range from 0.3 to several thousand nanometers.

The adsorptive properties of active carbon were utilized in antiquity. Hippocrates and his disciples recommended dusting wounds with powdered charcoal in order to remove their unpleasant odour. However, the rational use of active carbon for industrial purposes was started at the end of the 18th century. The Swedish chemist Karl Wilhelm Scheele, an apothecary by profession, was the first to discover (1773) the phenomenon of adsorption of gases on charcoal. A dozen years later (1785) the Russian academician Lovits from Saint Petersburg found that charcoal, when immersed in tartaric acid solution, decolorizes it by adsorbing the organic contaminants present. This discovery lead to the first industrial application of charcoal in the sugar industry in England in 1794, where it was used as a decolorizing agent for sugar syrup. This event initiated research on adsorption from the liquid phase. The discovery in 1903 of the selectivity of the adsorption process with respect to various substances in the liquid phase taking place on silica gel as adsorbent, is attributed to the Russian scientist Mikhail Tsvet, who was working at that time at the Technical University in Warsaw. He observed when passing an alcoholic solution of carotene through a layer of silica gel that regions of different coloration appear in the gel. Tsvet referred to this process as the adsorptive chromatographic separation of mixtures. This discovery was not only the beginning of a new analytical technique but also the origin of a new field of science.

In the years 1900–1901 R. V. Ostrejko was granted two patents which opened new prospects for the manufacturing technology of active carbon. The essence of the invention described in the first patent consisted in heating a vegetable material with metal chlorides, and that of the second patent in activating charcoal with carbon dioxide and steam at red heat.

In 1909 in Raciborz (Silesia), on the then German territory, a plant, called the Chemische Werke, was built — to manufacture for the first time on a commercial scale, powdered active carbon under the name of eponit, according to Ostrejko's patents. In 1911 the production was undertaken in this plant of new kinds of active carbon, known as norit and purit, obtained from peat by activation with steam. Following the Second World War this factory (now in Poland), became The Carbon Electrodes Works (ZEW). About the same time as the Raciborz factory was founded, a wood distillation plant was built in Hajnowka (East Poland), initially manufacturing active carbon solely from wood. The origins of the Norit factory in Amsterdam, now one of the most advanced international manufacturers of active carbon can be traced back to the same time. The process of the chemical activation of sawdust with zinc chloride was carried out for the first time in 1914 in the Austrian plant in Aussig, and also in 1915 in the dyestuff plant of Bayer. The powdered carbons were used at that time chiefly for decolorizing solutions in the chemical and food industries.

World War I introduced the tragic problem of protecting man's respiratory tracts from toxic warfare agents introduced intentionally into the air. In April 1915 in France and in May of the same year in the neighbourhood of Warsaw, the German army used warfare gases for the first time against the British and French in the West, and against Russian soldiers in the East. This gave rise to a hasty search for means of protection. Nikolai Zelinski, a professor of Moscow University, was the first to suggest the use of active carbon as the adsorption medium in gas masks. Such masks, of course with many modifications, are the basis for protecting the respiratory tracts of soldiers throughout the world to the present day. During World War I, coconut shells provided the raw material for the production of active carbon. These very tragic experiences and research conducted in the 1930's have led to the development of new technologies for obtaining granulated active carbons of supersorbon and benzosorbon type. These carbons have found commercial application in the adsorption of gases and vapours. The possibility of purifying city gas by removing benzene on active carbon, and other recuperative methods in which this adsorbent was used, have extended the commercial utilization of active carbon.

In the last two decades we have observed a rapid increase in demand for active carbon which is accompanied by an equally dynamic development of the science of the carbons themselves and of the adsorption phenomena proceeding on their surfaces. Active carbons are widely used as adsorbents of waste gases and vapours e.g. for removing CS_2 from air, of solvents for their recovery, of contaminants of aqueous solutions, e.g. the purification of sugar syrup, in the treatment of potable water and waste waters, in air conditioning devices, in vacuum technology, e.g. in sorption pumps, in adsorption gas chromatography, in the purification of blood and adsorption of toxins from systemic fluids, etc. Active carbons are also finding increasing application as catalyst supports as well as materials for electrodes in chemical sources of electricity. The ever-increasing hazards of environmental pollution open new important prospects for active carbons. Today the global annual production of active carbon amounts to about 500 000 tons, a large proportion of which comes from Europe and North America.

In this book, the production of active carbon is considered and its porous structure and chemical constitution of its surface are discussed. The methods of testing this carbon material are described and its adsorption properties are discussed. Finally some problems of application and regeneration of active carbon are reviewed. The number of technologies and raw materials used for this purpose being very large, the book concentrates on the most familiar, particular stress being laid on the chemistry of carbonization and activation processes which are of primary significance in the production of active carbon.

Since it is the porous structure and developed internal surface area with a complex chemical constitution that give rise to the specific properties of active carbon, these aspects are given particular consideration. As active carbon is used primarily as an adsorbent, one of the aims of the book is to present the adsorption processes occurring on the carbon surface. Simple, fundamental models of adsorption are presented, as well as the corresponding isotherm equations (or isotherms), beginning with the classical isotherms of Langmuir and Brunauer-Emmett-Teller, and ending with the contemporary model of adsorption on heterogeneous microporous solids. Special stress is laid on the theory of volume filling of micropores which, founded in the late 1940's, is still developing. Currently, we are witnessing a rapid extension of this theory. Many interesting reports have been published by numerous authors among which those of M. M. Dubinin (USSR), M. Jaroniec (Poland), K. S. W. Sing (Great Britain), and H. F. Stoeckli (Switzerland) deserve special attention. Also the present authors are deeply involved in the theoretical and experimental studies of the adsorption properties of active carbon and their modification, as well as in seeking new methods for describing these properties.

The final stage is the utilization of the fascinating properties of active carbon in industrial processes for purifying gases, separating mixtures, purifying solutions in the food industry, treatment of water and waste waters, etc. In the book significant space is therefore devoted to the applications of active carbon and to its regeneration, considered chiefly from the point of view of industrial practice. Original, and in the opinion of the present authors very interesting, results are also given of the medical applications of active carbon.

All the information given in the book, no matter how detailed, has the aim of aquainting the reader fully with all aspects of the manufacture, testing, and application of active carbon — this most popular adsorbent.

CHAPTER 2
Production of Active Carbons

2.1 GENERAL

Active carbon is obtained by the process of carbonization followed by activation of the initially carbonaceous material which is usually of vegetable origin. The product of the carbonization process, i.e. of pyrolysis of the carbon-containing material, conducted in the absence of air and any chemicals, is a material virtually inactive as regards adsorption, with a specific surface area of several $m^2\ g^{-1}$. Activation is necessary to convert this product into an adsorbent of high porosity with a strongly developed surface area. Both these processes, carbonization and activation, are the object of continued research since they are very important from the practical point of view for the demand for active carbons of different properties is ever increasing.

In the activation process the carbonaceous material is subjected to selective thermal treatment under suitable conditions which results in the formation of innumerable pores, fissures and cracks. The surface area occupied by pores per unit mass of the material increases significantly. Technological processes involve activation with chemicals or steam or gas. In chemical activation, raw materials not subject to carbonization are normally used, e.g. saw-dust and peat, whose mixture with inorganic activating reagents is subjected to high-temperature treatment. The activating substances are primarily dehydrating agents such as zinc chloride and phosphoric acid. In the steam-gas activation procedure such natural carbonized materials are used as: charcoal, peat coke, coconut shell carbon and cokes of hard or brown coal. The content of volatile substances in these materials is an important parameter of their susceptibility to activation. If this proportion is small as in graphite, activation is difficult or entirely impossible. When the content of volatile substances is high then activation yields an (approximately) proportional increase in reactivity of the activated substance. However, if the reactivity is too high, for instance in baking coals, the degree of activation can be lowered. The reactivity of the initial product is largely related to the occurrence of the largest pores (macropores).

Activation with gases involves usually the use of aerial oxygen, steam or carbon dioxide. Activation with air is selective but involves the risk of external (superficial) burn-off of the granules and consequently steam and carbon dioxide are preferred for commercial applications. To achieve a sufficiently high reaction rate when using the latter gases, the reaction must be conducted at 800–1000°C. This, however, requires special apparatus such as fluidized bed ovens, rotary ovens, multilevel ovens, fluidized bed reactors, etc. The choice of device depends on the grain size of the raw material used and on the required form of the carbon — as powder or extruded granules. Rotary ovens are readily available and therefore most commonly used commercially.

During activation, the mass of the carbonaceous material decreases significantly. Under optimal conditions this phenomenon is equivalent to increasing porosity. Thus it is possible to estimate to a first approximation the increase of carbon porosity by a gravimetric method. In this case it is convenient to determine the bulk density.

For the proper selection of an active carbon for a given application, the following parameters should be considered: size and composition of the granules, specific surface area of the pores, pore volume, pore volume distribution, and character and chemical structure of the carbon surface. According to their commercial form we may distinguish: powdered active carbons which are used as decolorants, crushed active carbons with irregular particles, and extruded (granulated) carbons which usually have the shape of cylindrical granules. In the manufacturing process the properties of active carbons can be modified by control of the raw material, activation procedure, and time and conditions of activation. However, the final properties of an active carbon depend on many factors. The total number of pores and their volume distribution depend mainly on the nature of the raw material used and on the physico-chemical parameters of the activation process. In the process of chemical activation of non-carbonized initial material, one obtains carbons of high adsorption capacity containing small pores. However, such a carbon is always contaminated with inorganic residues of the substances used in the activation process. If the same raw material (e.g. wood) is first subjected to pyrolysis and then activated with steam, one can obtain a product containing mostly small pores and free from additional impurities.

2.2 RAW MATERIALS FOR THE PRODUCTION OF ACTIVE CARBONS

The principal properties of manufactured active carbons depend on the type and properties of the raw material used. In Europe the most important raw materials used for this purpose are wood (sawdust), charcoal, peat, peat coke, certain types of hard and brown coal, and the semi-coke of brown coal. To produce active carbons, which should exhibit high adsorption capacity and a large volume of the smallest pores (micropores), coconut shells are usually used. In the USA, brown carbons and petroleum products are widely used for manufacturing active carbons. In the literature, possibilities are reported of using many other natural and synthetic materials for the purpose. Among these are the shells of various nuts, stones of many fruits, asphalt, metal carbides, carbon blacks, carbon-containing scrap-waste deposits from sewage, ash, scrap tyres, PVC and other polymer scrap [1]. These materials, however, have yet to find wider application in the commercial production of active carbon.

The greatest potential clearly lies with hard coals. Currently they account for 60% of the production of active carbons [2], and their role has increased significantly since 1945 [3]. Thus in Poland the production of extruded active carbons is based on hard coal. Crushed and powdered carbons are produced mostly from vegetable materials but also from hard coal. Wood is also a raw material of major importance. Active carbons with a strongly developed microporous structure (used in adsorption from the gas phase) are obtained on the commercial scale not only from hard coals (and also anthracite) but also from peat and wood. Carbons designed for adsorption from solution are produced mainly from peat, wood, brown coals, and the semi-coke from hard coal. The raw materials for producing particular types of active carbon, with the basic characteristics of their porous structure (and appropriate activation method) are listed in Table 2.1.

As already mentioned, coals play an important role in the production of carbon adsorbents. They are the source both of very cheap 'one-time' use adsorbents and of the more expensive granulated active carbons of high mechanical strength and developed micropore structure. The popularity of coals as raw materials is the ease with which they

Table 2.1
Examples of commercial and laboratory active carbons [4]

Symbol	Short description	Country of origin	Application	Refs.
1	2	3	4	5
Commercial active carbons				
AG-5	Commercial active carbon from Hajnówka, obtained from pit coal, activated with steam, size fraction *ca.* 1 mm	Poland	adsorption from gas phase	[4a]
Y-25	Product of Hokuetsu Tonso Corporation	Japan	adsorption of benzene and methanol	[4b]
BPL	Active carbon BPL-type: bituminous-coal base from Pittsburgh Activated Carbon Corporation. Surface area (BET N_2) 1050–1150 $m^2\,g^{-1}$, particle density 0.85 $g\,cm^{-3}$, pore volume 0.7 $cm^3\,g^{-1}$, ash max. 0.8 wt. %	USA	adsorption of phenol	[4c]
Merck	Active commercial charcoal; particle size 1.5 mm from Merck (Darmstadt)	FRG	adsorption from aqueous solution	[4d]
ASC	A standard commercial carbon with a similar grain size distribution, Calgon Corporation, Pittsburgh	USA	active carbon impregnated with copper and chromium	[4e]
F-400	Raw materials — bituminous coal, ash content 4–8%, Calgon Corporation, Pittsburgh	USA	water treatment	[4f]
Hydraffin 71	Lurgi-Bayer	FRG	adsorption from gas phase	[4g]
AX-21	A petroleum pitch-based active carbon of very high adsorption capacity manufactured and supplied by Anderson Development Co., Michigan	USA	adsorption from gas phase	[4h]
Carbosieve	A polymer molecular sieve carbon manufactured by Supelco and supplied by Bioscan, Canvey Island	USA	adsorption from gas phase	[4h]

Table 2.1 (continued)

1	2	3	4	5
SKT, AG	Soviet commercial active carbons	USSR	recovery of organic solvent vapours	[4i]
Saran B	Saran charcoal, a carbon produced by complete decomposition of poly(vinylidene chloride) (PVDC)-poly(vinyl chloride) (PVC) copolymer (Dow Chemical Company)	USA	commercial	[4j]
P10, P15, P20	Active carbon fibres (ACF) (Osaka Gas and Unicha). These pitch-based ACF were obtained through melting, spinning and activating pitches after the distillation of coal tar	Japan	adsorption of NO, SO_2, NH_3	[4k]
	Laboratory carbons			
C, M	Two botanically similar agricultural by-products: plum (series C) and peach (series D) stones have been used as precursors for the preparation of active carbons. Activation procedure used: carbonization in N_2 at 1073 or 1123 K for 2 h, followed by activation in CO_2 at 1073, 1098 or 1123 K for 8 h and 16 h	Spain	adsorption of N_2, CO_2, i-butane (gas-phase), p-nitrophenol and methylene blue (liquid-phase)	[4l]
D	Active carbons obtained from olive stones which had been carbonized at 1123 K in nitrogen and subsequently activated in following carbon dioxide at 1098 K	Spain	adsorption of N_2 and pre-adsorption of n-nonane	[4m]
CEP	Carbons of the CEP series resulting from activation of a soft wood precursor, CEP-0, with a very low ash content. Activation by CO_2 was carried out at 1123 K in a fluidized bed reactor, with burn-offs up to 80%	Switzerland	adsorption of CH_2Cl_2, C_6H_6, N_2 and CO_2	[4n]

Table 2.1 (continued)

1	2	3	4	5
A	Active carbon was prepared from chemically pure saccharose dissolved in a small quantity of water and heated at 570 K after evaporation of water. The resulting carbonizate was sieved to a size of 0.6–1.2 mm, heated under nitrogen at 970 K for 0.5 h. This coke (heated carbonizate) was activated at 720 K in oxygen in a rotating oven containing a 5 g sample of the carbonizate; oxygen flow rate 2.5 dm^3 h^{-1}, time 2 h	Poland	adsorption of argon, porosimetric measurement	[4o]
ACP	Sample was prepared from saccharose by treatment with calcium chloride at a salt/sugar ratio of 5:1. Maximum treatment temperature −1123 K. Product crushed and washed	USSR	adsorption of benzene vapour	[4i]

develop a porous structure. Coals exhibit an initial porous structure of their own. However, despite this, they cannot be used directly as commercial adsorbents since they have mostly very small pores which are inaccessible to the molecules of most adsorbates. To obtain active carbons suitable for commercial application, the coals require special treatment. Brown coal is the raw material featuring the greatest volume of pores. As the degree of metamorphism (content of elemental carbon) increases, the porosity of the carbonaceous material decreases [5, 6].

Various coals, from brown coal to the anthracites, may be used to produce carbon adsorbents. However, the manufacturing technology, price, properties and applications of the resultant active carbon depend on the type of initial material used. As regards processing, all coals fall into two groups. The first includes cherry coals and weakly baking coals; these are used for the production of cheap crushed carbon adsorbents. If granulated adsorbents are required, a bonding agent must be added. The second group includes baking coals which serve chiefly as the raw material for obtaining granulated adsorbents [2]. The mechanical strength of the granules depends in this case on the baking ability of the coal as a result of which it is possible to reduce or eliminate altogether the use of binding material, which leads to significant lowering of the costs of production.

2.2.1 Hard Coal

Hard coals are used for producing both granulated (extruded) and grain active carbons. The whole range of baking coals, from gas coal to coking coal, can be used in the production of active carbons. However, the granules of strongly baking coal form, when passing to the plastic state, a uniform coke mass. To avoid this it is necessary to decrease the superficial baking capacity of the granules, e.g. by their oxidation. The carbonized granules from coals of different degrees of metamorphism exhibit different reactivities in the activation process. The most active granules are obtained from gas coal, and the least active from coking coal.

The choice of coal as raw material for the production of active carbon should be made bearing in mind the required properties of the final product. The microporosity of coal increases with increasing metamorphism. Thus coals with a high degree of metamorphism constitute good raw material for the production of active carbons designed for adsorption of vapours and gases. Coals with a low degree of metamorphism and a high content of volatile matter are recommended for the production of active carbons with a wide distribution of pore volumes as a function of their radii [7].

To enable evaluation of a carbon for a given technological application, its physicochemical characterization is necessary, specifying (i) the type of coal, (ii) its purity, and (iii) its grading of size. We should note, however, that while the purity and grading of size can to some degree be modified, the type of coal has been determined by nature and therefore cannot be changed.

There are many national and international standards for classifying coals. In the present authors' opinion the most important are: ASTM Standard D388-84 (Standard classification of coals by rank), Deutsche Norme DIN 23003-1976 (Internationales Klassifikationssystem für Steinkohlen), and Norme Française NFM 10-001,1072 (Classification des houilles d'après leur nature). Here, in Table 2.2 we give the classification of coals according to the internationally recognised ASTM D388-84 standard which takes account among other things of the degree of metamorphism of the coal on passing from lignite to anthracite.

Table 2.2
Classification of coals by rank (ASTM D388—84)

Class		Fixed carbon limits, percent (dry, mineral-matter-free basis)		Volatile matter limits, percent (dry, mineral-matter-free basis)		Gross caloric value limits, Btu per pound (moist, mineral-matter-free basis)		Agglomerating character
		≥	<	>	≤	≥	<	
I Anthracitic	1. Meta-anthracite	98			2			non-agglomerating
	2. Anthracite	92	98	2	8			
	3. Semianthracite	86	92	8	14			
II Bituminous	1. Low volatile bituminous coal	78	86	14	22			usually agglomerating
	2. Medium volatile bituminous coal	69	78	22	31			
	3. High volatile A bituminous coal		69	31		14 000		
	4. High volatile B bituminous coal					13 000	14 000	agglomerating
	5. High volatile C bituminous coal					11 500	13 000	
						10 500	11 500	
III Sub-bituminous	1. Sub-bituminous A coal					10 500	11 500	non-agglomerating
	2. Sub-bituminous B coal					9 500	10 500	
	3. Sub-bituminous C coal					8 300	9 500	
IV Lignitic	1. Lignite A					6 300	8 300	
	2. Lignite B						6 300	

The Polish classification of coals according to Roga [8, 9] is based on the following properties characterizing the coals and their technological applicability: (i) volatile matter content, (ii) ease of baking (according to Roga), (iii) dilatometric properties (according to Arnu-Audibert), and (iv) heat of combustion.

When agglomerating or swelling carbons are used for the production of active carbon, the following technological scheme is usually applied [1]:

(1) grinding of the coal and spraying it with water,
(2) briquetting of the ground product,
(3) crushing the briquettes,
(4) sieve fractionation,
(5) oxidation to prevent swelling or baking,
(6) carbonization, and
(7) activation.

Oxidation is an important step. It proceeds with rapid evolution of large quantities of heat and is difficult to control since the temperature must be kept in a narrow interval. The industrial process is conducted in rotary ovens of fluidized-bed reactors. Heat is removed and the temperature controlled by means of the reaction gas or by spraying with water. The temperature is kept within the range 150–370°C depending on the kind of process and type of coal used. The oxygen content of the oxidizing gas can be varied within wide limits (1–50%). Carbonization is conducted at about 600°C and steam activation in the range 900–1000°C [1].

In the case of crushed coals with a high content of volatile matter and water, the material is first dried and then oxidized at 150–215°C by passing oxygen (1–3%) through the coal layer. The carbon is kept in contact with oxygen for about 19 h in the case of a stationary layer, and for 30 min in the case of a fluidized bed. Next the carbonaceous material is activated with oxidizing gases (steam, carbon dioxide, air) or activating chemical reagents, e.g. zinc chloride or phosphoric acid.

Coals of lowest grade, with a particularly high content of volatile matter, that coke with difficulty and yield grains of low mechanical strength, can also be used for the production of active carbons if they are first crushed and washed with a dilute solution of a mineral acid (hydrochloric, sulphuric or phosphoric). After drying, the grains are powdered, extruded with addition of bonding agents, carbonized and activated. It seems that the acid-treatment enhances processes which prevent the evolution of large quantities of volatile substances in the carbonization process, as a result of which active carbon granules of high mechanical strength are obtained.

Coals with a high ash content can be purified by flotation, wet oxidation, or other chemical procedures. For instance ash, which contains chiefly silicates, can be removed with solutions of sodium hydroxide or sodium carbonate.

In the USSR, extruded active carbons are produced from coals and coal-derived semi-cokes. Active carbons, mostly in crushed form, obtained from hard coals find wide application in the purification of water and waste waters. The Soviet literature recommends the use of active carbons obtained from anthracite for removing phenol from industrial waste waters [10]. Polish experience in this field shows that from baking coals one can obtain crushed active coals with a differentiated porous structure [11] which may be useful as cheap adsorbents for removing organic pollutants from coke-plant waste waters. The suitably-developed porous structure and high reactivity provide the basis for good activation properties of hard coals, and their baking capacity, lower evolution of volatile matter (as compared with brown coals) and the finer structure of the carbonizate ensure a higher mechanical strength of the carbon grains. The processes of obtaining crushed carbon adsorbents from hard coals have been developed on a commercial scale [12, 13].

The process of producing active carbons from anthracites, which reveal even in the untreated state significant adsorptive properties and high mechanical strength, attracts much interest. Anthracite and the coke obtained from it are characterized by a developed structure of micropores and lack of large open macropores. Anthracites exhibit, however, low reactivity which decreases further during thermal treatment, so their activation depends on the most efficient processes, especially in the fluidized bed [12, 14, 15]. In some cases the activation of anthracite is preceded by oxidation with aerial oxygen at low temperatures which increases the reactivity of the material [16]. A very active adsorbent can be obtained from anthracite by treatment with certain inorganic compounds ($NaOH$, Na_2CO_3, Na_2SO_4) [17]. The activity of anthracite can also be increased by treatment with nitric acid [18].

Many researchers have been working on the development of the technology of producing molecular sieves from anthracites [15, 16, 18]. It is known that hard coals including anthracite reveal molecular sieve properties which manifest themselves in a preferential adsorption of substances with small molecules. However, these properties have no practical consequence since the diffusion of a substance into the micropores of coal is very slow and thus the total sorptive capacity of anthracite is insignificant. Many attempts have been made to improve the adsorptive properties of anthracites by treating them with oxidizing gases [15, 18, 19]. It has been found, however, that activation of anthracite to a high degree of burn-off leads to an excessive expansion of the pores, as a consequence of which they lose their molecular sieve properties. Activated anthracites exhibit clear molecular sieve properties but are significantly inferior to zeolites.

2.2.2 Raw Materials with a Low Degree of Carbonization

Among the raw materials with a low degree of carbonization, brown coal, peat, wood-based materials and plastics are those of commercial importance. For instance in the USRR, brown coal constitutes the basis for production of cheap adsorbents for purifying water and waste waters, as well as for desulphurizing gases [2]. Brown coals are characterized by large pore volumes, from micropores to macropores, with a significant contribution of mesopores. It is assumed that for the purification of waste waters, active carbons (also granulated) can be used obtained from bituminous coals (active carbons with a high proportion of micropores for adsorption of small molecules) or lignite (active carbons with a high proportion of mesopores for adsorption of larger molecules) [20].

As a result of their developed porous structure, semi-cokes and cokes from brown coal are used as adsorbents. The branched system of transport pores enables easy access of the adsorbate molecules to the carbon surface, including large molecules adsorbed from solutions. Thus semi-cokes and cokes obtained from brown coals may be used as cheap grain carbons for one-time use in the treatment of waste waters [21–24]. By activation of brown coals, as a result of their high reactivity and developed structure of transport pores, uniformly activated narrow-porous adsorbents can be obtained. There are many methods of obtaining carbon adsorbents from brown coals [12–14, 25–31]. Processes have been developed on a semi-commercial scale of obtaining grain adsorbents from brown coals by activation in a fluidized bed [12, 14, 30, 31]. The active carbons obtained in this way feature a developed porous structure but their mechanical strength is insufficient; they have found application in the treatment of waste waters.

Activation of brown coals with gases in rotary ovens is conducted on a commercial scale in the USA. In West Germany a semi-coke from brown coal is produced that can be activated with gas without any preliminary treatment [1].

The disadvantage of almost all kinds of brown coal is their comparatively high sulphur content which, after activation, is present mostly as sulphites, resulting in the development of an unpleasant odour even in weakly acidic media. In many applications of this adsorbent it is necessary to remove this odour, which may be achieved by washing with acid. Another procedure consists of treating brown coal with hot water at 60–90°C in the presence of air. In view of its relatively low oxidation potential, the sulphur passes almost completely into solution (due of course to the catalytic effect of the activated carbon).

The high ash content in brown coal can often be reduced before carbonization and activation, e.g. by treatment with a mixture of oil and water, when the coal remains in the oil layer, while the ash passes into the aqueous phase. In this way the ash content can be reduced to 20% or even 10%.

Black peat is a particularly suitable raw material for obtaining active carbon. Its carbon content is about 60%, but the content of bonded carbon relative to dry matter is only 35%. In view of the high content of volatile matter, activation of peat with gases must be preceded by carbonization. Chemical activation can be carried out immediately after drying the peat.

Peat coke, obtained on a commercial scale in shaft ovens with external heating, e.g. to 850°C, is particularly suitable for gas activation. This coke contains almost 90% of bonded carbon, and its ash content is 2.5–4.5% [1]; it is easy to activate and the resulting products reveal specific surface areas of up to 1600 $m^2 g^{-1}$, as determined by the BET method.

Research on the production of active carbons from coal has shown that there are reasonable possibilities of controlling their porous structure [32]. If the degree of burn-off is small, the porous structure does not show micro- and mesopores; as the degree of burn-off increases, an increase in the proportion of micro- and mesopores is observed, followed by a distinct increase of the volume of the latter. If the burn-offs are high, the resulting active carbons show good decolorizing properties. Peat and wood are used as raw materials for the production of active carbons by the Norit company. Lignites are also traditionally used for manufacturing active carbons.

Charcoal is commonly used for the production of decolorizing active carbons, mainly in powdered form. It may also be extruded on the addition of bonding materials to obtain granulated active carbons. However, the carbons obtained in this way and activated, for instance, with steam have smaller specific surface areas as compared with the materials produced from raw materials with a high carbon content, and at the same time they reveal poorer mechanical strength [33]. Since the sources of wood as a raw material are shrinking, increasing interest is being concentrated on the possibilities of using waste materials such as sawdust or the waste shavings from the production of furfuryl alcohol from wood [8].

The charcoal currently used in the active carbon industry is for the most part not obtained from charcoal kilns. Thus the carbonization of wood is conducted in large-volume steel reactors. For this purpose such processes as the Deguss or Sific have been developed [1]. Finely ground wood waste, e.g. turnings or chips, can be carbonized in rotary or fluidized bed ovens. The grain or granulated carbons as well as pressed materials from powdered charcoal and bonding material are activated in shaft and rotary ovens with steam or carbon dioxide at 800–1000°C. The extruded or pelleted carbons containing a bonding material require preliminary heat treatment at about 500°C when the bonding material undergoes partial carbonization. The active carbons obtained from wood are noted for their high purity and large content of small pores. The carbonization of bark followed by activation with gases yields a cheap active carbon used for decoloration of industrial waste waters. In the laboratory an active carbon showing interesting adsorption properties was obtained after activation for only 0.5 h with steam at 870°C [34]. Attempts have been made to start commercial production of active carbon from bark.

In the literature it is reported that organic polymers can also be used as raw materials for the production of active carbons. The products are used as adsorbents with very large specific surface areas (e.g. active carbons of the Saran type) [35]. The possibility has also been considered of using synthetic resins as raw materials, e.g. phenol-formaldehyde ones, from which at low burn-offs active carbons of molecular sieve type are obtained [36, 37]. Considerable interest is devoted to fibrous active carbons which find application as air filters and respirators or as catalyst supports [38]. Most frequently fibrous carbon adsorbents are produced from rayon, polyacrylonitrile, PVC or Saran (PVC-vinylidene copolymer) fibre, lignin, pitch, etc.

From polyacrylonitrile or acrylonitrile copolymer textile wastes, nitrogen-containing active carbons can be obtained which exhibit very good adsorption towards mercaptans. These textile materials are initially carbonized in air at 500°C and then activated with steam at 950°C. In contrast to these carbons, those obtained from polyacrylonitrile at very high temperatures in a neutral gas atmosphere have a very low nitrogen content.

Activation of PVC scrap must be preceded by the removal of HCl by heating to 360°C in air; suitable activation involves treatment with steam at 800–1000°C. As a result, active carbon is obtained with a maximum specific surface area of 1300 m^2 g^{-1} and good adsorptivity with respect to methylene blue.

Research has been conducted on the possibilities of applying various kinds of industrial waste such as scrap rubber, bakelite or dumped gangue to the production of cheap active carbons [39].

The resources of raw materials used hitherto for producing active carbons being limited, the attempts to find new raw materials for this purpose are very important. For instance, the possibilities have been investigated of using olive stones [40, 41], plum stones [42], apricot and peach stones [43] for the production of active carbons. The use of shells of fruit stones as a raw material for producing active carbons yields adsorbents with very desirable properties [44] such as high homogeneity, significant hardness and resistance to abrasion, as well as a high volume of the micropores. The authors [40–44] claim that fruit stones obtained from fruit processing plants could provide a raw material for the production of cheap active carbons with a developed micropore structure.

Many manufacturers produce active carbon from coconut shells. Usually the shells are initially subjected to carbonization in rotary ovens, after which they are activated with steam. The granular active carbons obtained in this way feature high mechanical strength and very small pores. They find application primarily in gas masks and similar equipment. Active carbons with high mechanical strength can also be obtained from hazel-nut and fruit stone shells. Olive stones, which are a waste from the olive oil industry in the Mediterranean region, are also a raw material for the production of active carbon. These stones are initially treated with 10% sulphuric acid and with water, and then carbonized at about 800°C. The product has an internal specific surface area of about 500 m^2 g^{-1} which can be increased by activation to 1500 m^2 g^{-1} [40]. The oxygen content in these carbons is 3–5%.

In the USA active carbons are also obtained from peach stones. For some time American producers widely used waste pulp liquors and black ash residue. However, the production of active carbon from this waste gradually ceased as the use of hard coal and brown coal for this purpose increased.

In the literature mention is also made of the use of crude oil products, asphalt and carbon black for the manufacture of active carbons. In the USA granulated active carbons are obtained from the liquid fractions of crude oil. The coke obtained from heavy coal oil can be activated with steam at about 850°C. This process is conducted up to a 55% burn-off. Activation in a fluidized bed is conducted at 870°C for 10–13 h. Instead of steam, carbon dioxide or air can be used as the activating gas. The specific surface area of the coke activated in this way is 400–650 $m^2 g^{-1}$, and is near the lower limit of the values characteristic of typical active carbons. The use of these carbons is limited to the purification of waste waters.

Active carbons of greater specific surface areas can be obtained from salts of aromatic acids — products of petroleum coke oxidation with nitric acid. Also crude oil distillation residues and soft asphalt can be processed to active carbons. An extruded active carbon of high mechanical strength can be obtained from a mixture of asphalt and sulphur [1]. From furnace carbon black, it is possible to obtain a carbon featuring a very narrow pore volume distribution function [45]. In this case the surface of carbon black is coated with a thin layer of polymer which plays the role of a bonding agent in the carbonization process.

Active carbons can also be manufactured from tyre scrap [46] or bakelite waste [47, 48]. The latter waste, after grinding and screening, was carbonized in a pilot fluidized bed reactor. The resulting active carbons are suitable for use in water treatment and decolorizing of solutions. From bakelite waste, a granular active carbon was obtained which was used for adsorbing toxic compounds from the gas phase [49]. In the literature [50] a procedure has been described of producing active carbons from pine-tree and oak sawdust and pine-tree bark by a fluidized bed process.

This short survey of the raw materials used for producing active carbons allows us to conclude that the possibilities are quite broad. It should be noted, however, that only well-defined raw materials yield a final product with precise properties, and so the proper selection of the raw material for the production of active carbon is very important.

2.2.3 Binding Materials for Preparing Extruded Active Carbons

High-grade active carbons are obtained by preliminary mixing of the carbonaceous raw material with a binding agent, which usually is a wood or coal tar or else pitch. However, attempts have also been made to use resins and products from the petroleum industry for this purpose.

So far it has not been possible to provide clear principles for selecting a binding agent for extruding active carbon granules. According to Morgan and Fink [51], the binding material should melt to a large extent during coking, be a relatively small molecule, and exhibit a high coking temperature. The authors claim that dextrose and coal tar pitch are therefore typical binding agents, since the use of agents of low fluidity (e.g. urea-formaldehyde, polystyrene or ethylcellulose resins) leads to the production of active carbons of reduced mechanical strength. In commercial practice hard-wood tar is most frequently used, although other binding materials are also applied. The substances contained in tar that favour the agglomeration of the carbon particles also affect the development of the porous structure of active carbons and their mechanical strength. In view of the decreasing supplies of wood, one might expect that wood tars will become less available, and substitutes are being sought such as black oil, coke oven tar, generator tar, and asphalt [52]. Attention has also been drawn to the possibility of obtaining granulated active carbons without the use of binding agents by applying devices (plate granulators) yielding active carbons in the form of spherical granules. In this case water and waste sulphite liquors are added to the powdered raw material. The mechanical strength is ensured here as a result of the baking capacity of the hard coal used [2, 53]. The granules obtained from powdered coals and wood tar show high reactivity towards the activating agent, and the resulting active carbons have a fairly high mechanical strength. The use of coal tar leads to reduction of this reactivity but instead yields granules of better mechanical strength [2]. Thus it is possible to control the properties of these active carbons by varying the parameters of the binding material by using raw materials of different composition and physico-chemical properties. In Poland a technology has been developed of obtaining active carbons by producing pellets directly from baking coal powder. In this country modified wood tar has found common application in the commercial production of granulated active carbons. This tar plays the following functions in the process of extruding the coal-binding material paste:

(1) It binds the single particles of the carbon material together and imparts to the extruded agglomerate suitable mechanical strength that allows its further processing.

(2) Gives the granulate, after carbonization, resistance to crumbling and abrasion.

(3) Produces in the process of physico-chemical activation, free intergrain spaces facilitating the development of specific shapes and sizes [54]. Research is under way to find new bonding agents. Attempts have been made to use waste sulphite liquors, brown coal tar, sodium carboxymethylcellulose, gas pitch [55], cumene tar and asphalt [54].

2.3 TECHNOLOGICAL OPERATIONS IN THE PRODUCTION OF ACTIVE CARBONS

2.3.1 Granulation

This process involves the preparation of a paste composed of finely ground coal dust and tar, extrusion of the granules and drying (possibly with superficial oxidation). In the course of thermal treatment, the binding agent reacts with the carbonaceous material to yield a resistant spatial polymer. The most common method of producing cylindrical granules consists of extruding the plastic paste including the coal-binding agent through a nozzle. The filaments obtained are cut into cylindrical granules of required dimensions. The mechanical strength of the granules depends on the mechanical properties of the initial carbonaceous material and on the granulation method applied. Particularly important are, however, the physico-chemical properties of the coal and binding agent. The greater the similarity of these components, the stronger are the bonds between them [56]. In view of this, coal tars should be the most appropriate binding agents for hard coals. Indeed, by using coal tars or pitches, granules of high mechanical strength are obtained. However, the mechanical strength of the granules is not the only important feature, their chemical reactivity ensuring the generation of a developed porous structure in the activation stage also being essential. Thus the use of coal tar, which features low reactivity, presents a number of problems. For this reason the use of wood tars as binding agents is most common [57, 58]; the latter ensure sufficient mechanical strength and high reactivity of the granules.

Many methods of granulating active carbons have been developed that allow the production of spherical granules [59, 60]. Such granules have considerable advantages over cylindrical or irregularly shaped ones, the latter being particularly liable to crumbling or abrasion. The spherical granules having no sharp edges nor protrusions and, revealing a high mechanical strength, are not liable to abrasion. Furthermore, the packing of spherical grains guarantees the highest ordering: the channels between the spherical granules have regular shapes so the layers of such granules show a lower resistance compared with cylindrical or irregularly shaped ones. If we assume that the resistance of a layer of spherical granules to the passing medium (gas or liquid) is unity, then that of cylindrical and irregular granules is 2.5 and 3.3, respectively [59]. In the USA a method has been developed of obtaining spherical granules in drum granulators [59, 60]. This process consists: (i) in grinding the initial coal, mixing it with the ground, dry bonding agent, which in this case may be pitches with a softening point ranging from 80 to 150°C or petroleum bitumens, and (ii) in granulation in drum granulators with addition of water in a proportion of 30–50% of the dry coal. Granulation in drum granulators is very efficient but the resulting granules are far from ideally spherical, and these granulators also make it difficult to obtain granules in a narrow size distribution [61].

In Japan and the USSR, processes have been developed of granulating coals by means of disc granulators. The efficiency of these devices is only slightly inferior to that of the drum granulators but the granules obtained have a regular spherical shape and, as a result of the fractioning action of these granulators, the granules can be obtained in fairly narrow-size fractions. By varying appropriately the rotational speed of the discs, the height of their rim, and their horizontal tilt, granules of the requisite size are obtained.

Baking coals are gaining importance as raw materials for producing granulated adsorbents as they permit elimination of the expensive binding agents and at the same time simplification of the technological process. The important characteristic of the baking coals is that on heating to a given temperature, they assume a plastic state of greater or lesser fluidity. Further heating leads to the solidification of the mass due to polycondensation. If sufficient contact between the coal particles is ensured then they sinter, yielding a conglomerate of good mechanical strength. The transition of the coal into the plastic state manifests itself in the evolution of volatile products which produce a swelling of the mass which, after hardening, preserves the porous structure. The character of the porous structure and the mechanical strength of the granules depend directly on the properties of the plastic mass (its viscosity and mechanical strength) which in turn depend on the properties of the initial coal and on the conditions of its treatment.

In the USSR a method has been developed based on baking coals of producing carbon adsorbents in the form of spherical granules [62]. Also in the USSR a one-step process has been developed of obtaining spherical granules in a disc granulator 1.5 m in diameter without adding bonding agents. The consumption of water is 20% per mass of dry coal, and the bulk weight of the dried granules is 0.60–0.62 g cm^{-3}.

In Japan spherical granules 1–7 mm in diameter have been obtained by mixing crushed coal with water and the binding agent, following which the paste was fluidized and granulated in a disc granulator of a disc diameter 1 m [63]. Spent sulphite liquor was used as the binding agent. The consumption of bonding agent is in this case 15% and that of water 6–15% per unit mass of crushed coal.

2.3.2 Carbonization

This is one of the most important steps in the production process of active carbons since it is in the course of carbonization that the initial porous structure is formed. Activation merely develops this structure further in the same direction, since any significant change by means of a suitable choice of activation parameters is usually impossible.

The granules formed from powdered coal and the tar binding material are first dried at about 200°C to remove the excess of lower boiling fractions of the binding agent and to give them the mechanical strength required in the further operations. The dried granules are carbonized at temperatures up to about 800°C (usually at 600°C).

Preliminary oxidation of the granules has a significant effect on their properties. Slight oxidation prior to carbonization enhances the mechanical strength and density of the carbonized granules. This is due to a somewhat reduced baking of the oxidized coal and a resulting reduced swelling of the granules. However, excessive reduction in the baking ability of coal due to prolonged oxidation of the granules gives the opposite effect, i.e. it decreases their strength and density as compared with the optimal regime.

The carbonaceous material that constitutes the basis for the manufacture of active carbon by the steam-gas method must meet certain requirements among which the most important are: (i) low content of volatile matter, (ii) high content of elemental carbon, (iii) a definite porosity and (iv) sufficient strength of attrition. Of course, natural carbonaceous materials do not meet all these requirements simultaneously and therefore they require carbonization [64]. However, even if a carbonaceous material did meet all the requirements mentioned it would require carbonization after the ground carbon material had been extruded with a tarry binding material to obtain granules, since the bonding material must be converted to coke that bonds the particular grains. Under such conditions the carbonaceous material also undergoes thermal decomposition. If the carbonaceous material had been pre-carbonized, e.g. as charcoal or semi-coke from brown or bituminous coal, then the chemical changes taking place in the binding material are made different because of the altered physicochemical character of the surface and porosity of the semi-coke used for granulation. Were we to subject a raw material of low carbon content directly to activation, it would nevertheless undergo carbonization in the course of the gradual increase of temperature and simultaneous interaction with the activator. However, such a procedure would not be conducive to realising a product of the expected porosity, since the above-mentioned processes would then overlap and the product obtained would have a random porous structure dependent on which factor — rate of increase of temperature or chemical activity and activator concentration — prevailed at a given moment. Carbonizates of the requisite properties are obtained by suitable adjusting the conditions under which the pyrolysis of the carbonaceous material is carried out.

As mentioned earlier, the carbonization process proceeds at 500–800°C. The granules acquire mechanical strength in this process, and due to the evolution of volatile matter, the material becomes richer in carbon and the initial porous structure develops. The heat destroys the organic matter of the initial coal as well as of the binding material. The homogeneous granule is formed as a result of polycondensation and polymerization. The initial porous structure of the coal undergoes significant changes during thermal treatment. In non-agglomerating coals, a gradual increase of the volume of micropores takes place up to 600–700°C. Further heating leads to compression of the material and decrease of the pore volume. In the case of agglomerating coals, the process proceeds in a more complex way, however, in that during the principal stage of heating we also observe an increase in the pore volume; the coal passes to the plastic state and its initial porous structure is destroyed. Further heating produces the evolution of volatile substances from the solidified plastic mass as a result of which a branched system of pores is formed. The later stages proceed in a similar manner as for non-agglomerating coals.

Carbonizates of requisite properties are obtained by suitable adjustment of the conditions of pyrolysis of the bituminous material. These relevant parameters are: (i) the final temperature achieved, (ii) the time of carbonization, (iii) the rate of temperature increase, and (iv) the atmosphere in which the pyrolysis is conducted. The most important of these parameters is the final temperature of the process; this is associated with the need to supply to the macromolecules of coal significant amounts of energy in order to produce splitting of the weaker chemical bonds and to enable migration of the volatile products of thermal decomposition of the raw material to the granule or grain environment. Additionally, in the solid product of pyrolysis, i.e. in the carbonizate, some ordering of the compact carbon matter — much greater than that of the initial coal — can take place, and this may occur only at an elevated temperature. At high temperatures, bituminous raw materials behave in diverse ways. Those that have by nature a more ordered chemical structure, e.g. coals of higher grades, shrink with increase of the carbonization temperature as a result of which the total volume of the finest pores formed in the initial stage of carbonization decreases quite significantly. Moreover the reactivity of the carbonizates obtained after pyrolysis at 300°C is lower than that of carbonizates obtained at 600°C [64]. The decrease of reactivity becomes greater, as the carbon content of the coal decreases, i.e. with the greater intensity of pyrolysis as the temperature increases. It should be noted that as the carbonization temperature is increased, the condensation processes in the material are enhanced and the greater becomes the mechanical strength of the

Table 2.3
Effect of the final temperature of carbonization on the properties of granules
(after Kostomarova and Surinova [62])

Final carbonization temperature /°C	Mechanical strength of the granules /%	Bulk density of granules /g cm^{-3}	Rate of oxidation with CO_2 at 820°C /g(g min)$^{-1}$	Activation energy of the granules /kJ mol^{-1}
400	unstable	0.62	—	—
500	72.7	0.60	—	—
600	93.8	0.61	0.00194	212
700	98.0	0.65	0.00156	237
800	98.4	0.69	0.00134	256

3 — Active Carbon

resulting granules. The condensation of the material associated with the decrease of its pore volume results in a decrease of reactivity of the granules in the succeeding activation process (see Table 2.3). The volume of the micropores is in this case so small as to be inaccessible to benzene molecules.

The residence time of the bituminous material at the final carbonization temperature has an effect on the ordering of the compact structure of the carbon material. We distinguish here two principal temperatures at which the effect of time is different:

(i) A temperature lower than that at which the main thermal decomposition reactions appropriate to the given raw material are terminated.

(ii) A temperature higher than that of the internal transformations at which the final porous structure of the carbonizate is established.

In the first case, a further though slow thermal decomposition of the carbon material continues over time. However, some part of the decomposition processes has become inhibited. The carbonizate obtained under these conditions shows a greater reactivity towards the activating agents than that in which the reactions of the volatile pyrolysis products between each other and the carbon material, the ordering of the chemical structure of the carbon material, and the generation of carbon crystallites have been brought to an end.

In the second case, i.e. at a temperature higher than that at which the main thermal decomposition processes are terminated, with elapse of time a further ordering of the internal structure of the carbon material proceeds with the possible generation of crystallites. As carbonization continues, the volume of the smallest pores usually decreases due to the further decrease of the volume of the carbonized mass. For the reasons described, the reactivity of the carbonizate obtained becomes lower the longer it is maintained at the final carbonization temperature.

An important parameter of the carbonization process is the rate at which the final temperature is achieved. When the temperature is raised rapidly, the particular phases of the thermal decomposition of coal and the secondary reactions of the pyrolysis products with each other overlap, so control of the establishment of the porous structure in the carbonizate is more difficult. Neither does a high rate of increase of temperature favour higher ordering of the chemical structure of the carbon material, since the processes taking place in the solid phase proceed fairly slowly and therefore require time. If the temperature is raised rapidly, a large quantity of volatile matter evolves within a short

Table 2.4
Effect of the rate of heating on the properties of carbonized granules
(after Kostomarova and Surinova [62])

Heating rate /°C min^{-1}	Mechanical strength of granules /%	Bulk density of granules /g cm^{-3}	Pore volume of granules/cm^3 g^{-1}			
			total	micro-pores	meso-pores	macro-pores
1	93.0	0.71	0.24	—	—	—
5	94.0	0.70	0.22	0.12	0.01	0.09
8	94.1	0.72	0.23	—	—	—
20	91.5	0.62	0.28	—	—	—
stepwise growth	60.0	0.28	1.09	0.11	0.03	0.95

time, and as a result pores of greater sizes are usually formed. The reactivity of the carbonizates obtained in this way is greater than that of the products heated at a slow rate. This is due to the greater porosity and reduced ordering of the compact carbon material as compared with carbonizates obtained from the same raw material but at a low rate of heating.

The significance of the rate of heating in carbonization on the properties of the granules is evident from the fact that from the same raw material we can obtain granules with a total pore volume of 0.2 to 1.1 cm^3 g^{-1} (Table 2.4). The most resistant granules are obtained at a fairly low heating rate — up to 10°C min^{-1}. An increase of the heating rate to 20°C min^{-1} leads to a certain increase of the total pore volume but at the same time to a lowering of the mechanical strength of the granules.

The thermal decomposition of coal and the course of the secondary mutual reactions of the pyrolysis products and of the reactions of the latter with the solid carbonizate are also affected by the atmosphere in which the carbonization process is conducted. If the gases and vapours evolving during pyrolysis are rapidly removed by a neutral gas or combustion gases, then the quantity of the carbonizate obtained is smaller but its reactivity is greater. This is clearly evident when the combustion gases contain large amounts of water vapour and carbon dioxide, since then they react additionally with the decomposing coal even at relatively low temperatures.

The main aim of the carbonization process is to generate in the granules and grains the required porosity and ordering of structure of the compact carbon material. This ordering should be much greater than that of the raw material used. Both these factors have a crucial effect on the reactivity of the carbonizate in its reaction with the gaseous activating agent. This reactivity increases (i) with the degree of porosity generated and (ii) with reduction in the ordering of the compact carbon matter. A large volume of pores in the carbonizate facilitates the diffusion of the gaseous activator into the granules and ensures a large surface area on which chemical reactions may take place.

Usually the carbonization of granules is conducted in one of the following three types of device: stationary ovens with heat transfer through the walls, rotary ovens with internal or external heating, and fluidized bed ovens. During the carbonization process in stationary conditions, a temperature gradient appears between the region neighbouring the wall and the central part of the oven as a result of which the obtained carbonizate is qualitatively inhomogeneous. Rotary ovens with internal heating have a simple construction and therefore are very cheap. However, the combustion gases used as the heating medium contain oxygen which causes significant losses of the carbonaceous material due to its partial burn-off. A considerable decrease of the mechanical strength of the granules also ensues. Therefore the commonest process of carbonization is that in rotary ovens with external heating under a neutral atmosphere. If the heating rate is small, granules of increased mechanical strength are obtained. The rotation of the oven creates conditions favourable to the complete and uniform carbonization of the material.

Contemporary research in this technology is focused on carbonization of granules in a fluidized bed. Fluidized bed ovens have a high efficiency, but their disadvantage lies in the high rate of heating leading to swelling of the granules which hinders the production of a material with high strength and precisely determined properties. In some cases the carbonization of the granules is conducted in the presence of air [27, 65]. In order to lower energy consumption, the heat of combustion of the volatile products is utilized in heating the fluidized bed oven.

In current industrial practice, the carbonization of dried coal-pitch granules is usually conducted up to about 600°C at a moderate heating rate of 100 to 300°C h^{-1} [64].

In Poland research continues on the fluidized bed technology of producing active carbons. A block diagram for the production of active carbon bituminous coal by the fluidized bed method is shown in Fig. 2.1. The coal, ground to a grain size of 0–4 mm, is subjected to thermal

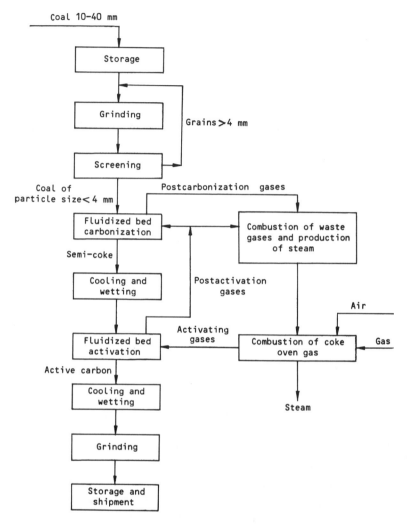

Fig. 2.1 Block diagram of production of active carbon from coal by the fluidized bed method (after Haberski and Heilpern [66]).

treatment in two separate steps conducted in two different fluidizedbed reactors. In the first, carbonization step the coal is heated to 500–550°C when tar and heavy hydrocarbons are removed, and the grains become highly porous. In the second, activation step the carbonizate is activated in a fluidized bed at about 1000°C. A mixture of steam and combustion gases constitutes the dynamic medium keeping the fluidized bed in suspension as well as the heating medium. The two steps considered, carbonization and activation, allow us to obtain from bituminous coal an adsorbent of moderate adsorptive capacity.

Studies on the utility of fluidized bed methods for obtaining extruded active carbons have shown that the quality of the adsorbents intended for adsorption from the gas phase is considerably affected by the way in which the carbonaceous material is degassed in the carbonization process. It has also been found [67] that the different porous structures obtained from different treatment regimes in the carbonization and activation processes have no significant effect if the adsorbent is used for adsorption from the liquid phase.

The fluidized bed methods of producing active carbon described in the literature allow us to draw the following conclusions:

(i) The fluidized bed carbonization and activation methods are suitable for the manufacture of extruded active carbons, especially of those intended for adsorption from the liquid phase.

(ii) The adsorbents obtained by these methods exhibit fairly good decolorizing properties, comparable with active carbons obtained by conventional methods from vegetable-based raw materials.

2.3.3 Activation

Now the product of carbonization has a weakly developed porous structure, so without additional activation it cannot be used in practice as an adsorbent. The basic method of activating coal-based granules consists of their treatment with oxidizing gases (steam, carbon dioxide, oxygen) at elevated temperatures. In the activation process, carbon reacts with the oxidizing agent and the resulting carbon oxides diffuse from the carbon surface. Owing to the partial gasification of the granules or grains, a porous structure builds up inside them. The structure of the carbonization product consists of a system of crystallites similar to those of graphite bonded by aliphatic-type bonds to yield a spatial polymer. The spaces between the neighbouring crystallites constitute the primary porous structure of the carbon. The pores of the carbonized granules are often filled with tar decomposition products and are blocked with amorphous carbon [68]. This amorphous carbon reacts in the initial oxidation step, and as a result the closed pores open and new ones are formed. In the process of further oxidation, the carbon of the elementary crystallites enters into reaction due to which the existing pores widen. Deep oxidation leads to a reduction in the total volume of micropores due to the burning off of the walls between the neighbouring pores, and in consequence the adsorptive properties and mechanical strength of the material decrease. Carbon oxidation is a complex heterogeneous process encompassing the transport of reagents to the surface of the particles, their diffusion into the pores, chemisorption on the pore surface, reaction with carbon, desorption of the reaction products, and diffusion of these products to the particle surface. The concentration profile of the oxidizing agent in the granule volume, and hence the formation of the carbon porous structure, depends on the rates of the particular steps of the process. However, temperature has a great effect on the oxidation process. At low temperatures the rate of the chemical reaction of carbon with the oxidizing agent is small, so it is this reaction that limits the overall rate of the process. This results in a dynamic equilibrium becoming established between the concentration of the oxidizing agent in the pores and that in the interparticle spaces. In such a case the activation process yields a homogeneous product with a uniform distribution of the pores throughout the whole volume of the granule. With increase of the oxidation temperature, the rate of the chemical reaction increases much faster than that of diffusion, and then the overall rate of the process becomes limited by the rate of transport of the oxidizing agent into the granule. At very high temperatures the oxidation reaction rate becomes so high that the whole oxidizing agent reacts with carbon on the external surface of the granule. In such a case significant losses of the material occur due to superficial burn-off, and a porous structure is not formed.

The rate of the oxidation process is limited by the reactivity of the initial carbonaceous material towards the oxidizing agent. The greater is the reactivity of the substrates, the lower the optimal temperature of the process at which uniform formation of pores in the granule is observed. The greatest reactivity is observed in the case of brown coal cokes, and the lowest reactivity in the case of anthracite coke. The bituminous coal cokes have an intermediate reactivity. Hence activation of brown-coal based carbons is conducted at relatively low temperatures and that of anthracite-based ones at relatively high temperatures.

Among the oxidizing agents listed above, oxygen shows the greatest and carbon dioxide the least activity.

The reaction of carbon with oxygen yields simultaneously carbon oxide and carbon dioxide according to the following stoichiometric equations:

$$C + O_2 \rightarrow CO_2 \qquad \Delta H = -387 \text{ kJ mol}^{-1} \qquad (2.1)$$

$$2C + O_2 \rightarrow 2CO \qquad \Delta H = -226 \text{ kJ mol}^{-1} \qquad (2.2)$$

i.e. both reactions are exothermic. Their mechanism, however, is not yet fully understood. It remains to be resolved whether carbon dioxide is the primary product of carbon oxidation or the monoxide is formed first and the dioxide is the product of secondary reaction. Current views are that both oxides are primary products [57], and the CO/CO_2 ratio increases with increase of temperature [69]. A carbon activated with aerial oxygen contains many oxygen-containing functional groups on the surface. Nevertheless aerial oxygen is very seldom used as an activating agent since it involves many difficulties. The reactions with oxygen being exothermic it is not easy to keep the temperature in the oven at the required level, the more so as oxygen is a very aggressive reagent.

The basic reaction of water vapour with carbon is endothermic and the stoichiometric equation has the form:

$$C + H_2O \rightarrow H_2 + CO \qquad \Delta H = +130 \text{ kJ mol}^{-1} \qquad (2.3)$$

This process has been studied extensively since it dominates not only the activation reaction but also the production of water-gas. The reaction of carbon with water vapour is inhibited by the presence of hydrogen while the influence of carbon monoxide is practically insignificant. The rate of gasification of carbon by water vapour is given by the formula [70]:

$$v = \frac{k_1 p_{H_2O}}{1 + k_2 p_{H_2O} + k_3 p_{H_2}} \qquad (2.4)$$

where: p_{H_2O} and p_{H_2} are the partial pressures of water and hydrogen, respectively, k_1, k_2, k_3 are the experimentally determined rate constants. The mechanism of reaction of carbon with water vapour can be presented with reasonable confidence by the following set of equations:

$$C + H_2O \rightleftharpoons C(H_2O) \qquad (2.5)$$

$$C(H_2O) \rightarrow H_2 + C(O) \qquad (2.6)$$

$$C(O) \rightarrow CO \qquad (2.7)$$

The inhibiting effect due to hydrogen may be assigned to blocking of the active centres by its adsorption.

$$C + H_2 \rightleftharpoons C(H_2) \qquad (2.8)$$

According to Long and Sykes [71] in the first step of the reaction the adsorbed water molecules dissociate according to the scheme:

$$2C + H_2O \rightarrow C(H) + C(OH) \qquad (2.9)$$

$$C(H) + C(OH) \rightarrow C(H_2) + C(O) \qquad (2.10)$$

Hydrogen and oxygen are adsorbed at neighbouring active sites which account for about 2% of the surface area.

The reaction of carbon and water vapour is accompanied by the secondary reaction of carbon monoxide with water vapour (the so-called homogeneous water-gas reaction) catalysed by the carbon surface:

$$CO + H_2O \rightarrow CO_2 + H_2 \qquad \Delta H = -42 \text{ kJ mol}^{-1} \qquad (2.11)$$

Reif [72] explained the presence of carbon dioxide and the catalytic surface effect of carbon by the following reaction:

$$CO + C(O) \rightleftharpoons CO_2 + C \qquad (2.12)$$

Steam activation is conducted at 750–950°C (Fig. 2.2). It is catalysed by alkali metals, iron, copper, oxides and carbonates, etc. The rate of carbon gasification with carbon dioxide is described by an expression analogous to that for the reaction with steam:

$$v = \frac{k_1 p_{CO_2}}{1 + k_2 p_{CO} + k_3 p_{CO_2}} \qquad (2.13)$$

where p_{CO_2} and p_{CO} are the partial pressures, and k_1, k_2, k_3 are the experimentally determined rate constants.

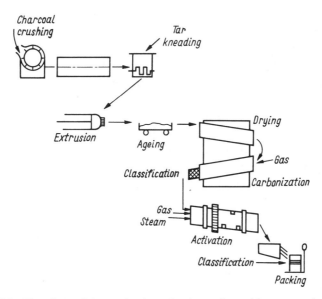

Fig. 2.2 Flowsheet of the production of active carbon with steam activation (after Smíšek and Černý [57]).

Two different mechanisms of interaction of carbon dioxide with the carbon surface are proposed:

(1)

$$C + CO_2 \rightarrow C(O) + CO \qquad (2.14)$$

$$C(O) \rightarrow CO \qquad (2.15)$$

$$CO + C \rightarrow C(CO) \qquad (2.16)$$

(2)

$$C + CO_2 \rightarrow C(O) + CO \qquad (2.17)$$

$$C(O) \rightarrow CO \qquad (2.18)$$

The basic difference between these two schemes lies in the different explanations of the inhibiting effect of carbon monoxide.

The reactions of carbon with water vapour and carbon dioxide proceed with absorption of heat, the reaction rate of carbon with carbon dioxide being, at a given temperature, 30% lower than that with water vapour [68, 73]. The use of oxygen as activating agent presents particular difficulties which are due to its exothermic reaction with carbon. In this case it is difficult to avoid local overheating in the activation process. In view of its high rate, the carbon burn-off process proceeds chiefly on the surface of the granules, producing high losses of material. In many processes, oxygen activation is conducted at very low temperatures and combined with treatment with water vapour. Such a method is most convenient when materials of low reactivity are activated.

The activation of carbon by means of oxidizing agents is conducted in gas ovens of varied construction and mode of operation. These ovens may be classified as shaft, rotary and fluidized bed.

Shaft ovens possess a system of chamber (known as 'muffles') arranged one above the other and filled with the material to be activated. The walls of the chambers are lined with roof-shaped bricks. The carbon material to be activated is loaded from the top, and steam is introduced from beneath. The carbon material moves by gravity down the shaft in countercurrent to superheated steam. Vertically three sections may be distinguished in which carbonization, activation and cooling take place successively. The residence time of the carbonaceous material is 2.5–3.0 tons per day. The incorporation in the ovens of fillings or deflectors increases the reaction surface area and facilitates mixing of the material.

Sec. 2.3] **Technological operations** 43

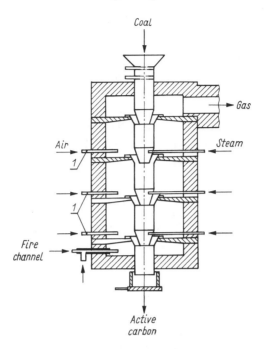

Fig. 2.3 Scheme of a shaft oven (after Kienle and Bader [1]).

In Fig. 2.3 a scheme is shown of a multistep oven with side furnaces and channels admitting the reaction gases. The popularity of these ovens is due to their energetic self-sufficiency. Heating of these ovens is necessary only at the start, since further heating is conducted via the heat of combustion of the gaseous products of carbonization. The drawbacks of these ovens are their high cost, low efficiency and non-uniform activation of the material due to the flow of the gaseous activator through the multiparticle-size layer of this material. As it has been shown experimentally by Dziubalski, Korta and Lasoń [74], the burn-off rate of the particular uniparticle-size layers of the carbonizate in the case of activation of a multiparticle-size layer is greatest in the lower layer and smallest in the upper layer. Thus the product of activation of a multi-particle-size layer of carbonizate is not a single carbon material, but at least as many different products as there were uniparticle-size layers in the multiparticle-size layer.

Better contact of the activating gas with the activated material and a more uniform temperature distribution in the carbon layer are achieved in rotary ovens [75, 76]. Such ovens are used to activate fine porous particles or extruded carbon materials. The contact between the carbonaceous material and the activating gases may be considerably improved by applying mixing devices. The time of activation depends on inclination angle of the oven and also on whether deflectors are present. The material to be activated and the gases may be introduced in co- or countercurrent. Two main types of design of ovens are distinguished, i.e. those with internal and those with external heating. The rotary ovens with internal heating are provided in their upper part, where the carbon material is fed, with a burner for fluid fuel. In this case the inside of the oven is lined with roof-shaped brick. To facilitate exchange of heat, openings allowing admission of air and the burners for internal heating are located in the bottom part of the oven. The pipe connections for admitting steam are usually located at the upper end of the oven. They may be mobile when steam is to be supplied to the surface of the carbon at different angles, which enables the production of active carbons with various pore dimensions. A scheme of a rotary oven with internal heating is shown in Fig. 2.4. Such ovens are heated directly by combustion of coke-oven gas, producer gas or petroleum. The advantage of rotary ovens is their low cost, their disadvantage is the excessive movement of the carbonaceous material, which causes abrasion by crushing. Moreover, these ovens are large and heavy which makes adequate technological control difficult.

More recently fluidized bed ovens have become quite common [65, 77–79]. In these ovens every particle of the material is in constant motion and in direct contact with the oxidizing gas present in excess, whose concentration is then constant throughout the whole volume. In the fluidized bed regime the temperature can be accurately controlled in the activation zone as a result of which the overall process can carefully be managed. In these ovens the activation time is much shorer compared with ovens of alternative design. A fluidized bed reactor of simple construction consists of a hermetic cylindrical or cuboid chamber provided in the lower part with a perforated plate through which the reaction gases are admitted. The process can be conducted batch-wise or continuosly. Multistep reactors are also known consisting of several vertically or laterally connected chambers, and also reactors made up of a number of sections separated by partitions. The fluidized bed ovens are used for activating fine-particle materials, and in particular cases, of extruded carbon.

The start-up of a fluidized bed reactor is usually rather difficult. Again, the fluidized bed process always involves the danger of ejection of carbon dust from the oven due to the considerable disintegration of the

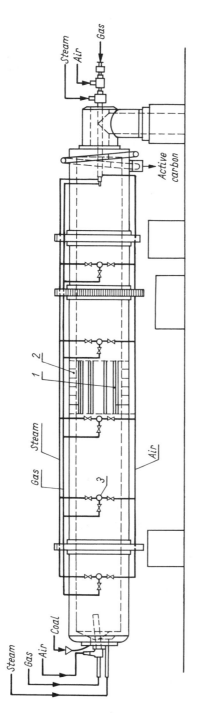
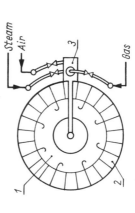

Fig. 2.4 Scheme of a rotary oven: 1 — mobile deflecting vanes, 2 — roofing brick lining, 3 — burner (after Kienle and Bäder [1]).

material. If the heating gases are introduced through a perforated plate, then their high temperature, indispensable for the success of the activation process, may induce the agglomeration of ash particles which are then deposited on the perforated plate and clog its meshes, thereby disturbing the uniform flow of the gas. As a result the local uniformity of the 'boiling' bed may be affected, abrasion of the coal particles becomes excessive, and coal dust may be ejected from the reactor.

The process may be improved by direct heating of the internal space of the reactor using the heat obtained from the combustion of carbon monoxide and hydrogen generated in the activation process of the carbonaceous material with steam. However, in this case accurate control of the amount of oxygen introduced into the reactor is crucial to prevent excessive surface burn-off of the particles and significant losses of the final product. Another possibility of increasing the yield lies in external heating of the reactor. In Fig. 2.5 a scheme of such an oven is

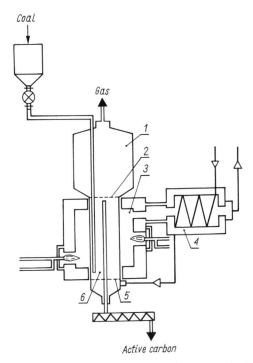

Fig. 2.5 Scheme of a fluidized bed reactor for gas activation of carbon: 1 — fluidized bed space, 2 — lower level of the fluidized bed, 3 — external heating space, 4 — heat exchanger, 5 — perforated separation plate, 6 — the reactor (after Kienle and Bader [1]).

given in which the heated activating gases flow with a velocity that affects neither the stability of the lower layer nor the fluidized bed condition of the upper layer. The separation of the external heating of the oven from that of the fluidized bed enables independent control of the heating and of the gas flow velocity in the fluidized bed. This in turn enables gentle activation of various carbonaceous materials. An advantage of ovens of this kind is their compact structure and homogenous activation. Among the disadvantages of fluidized bed ovens are the large consumption of gas, abrasion of the material being activated, and the exacting requirements as regards the composition and particle size of the treated granulate, as well as the agglomeration of the carbonaceous material.

Improvements in the design of fluidized bed ovens are directed towards the development of continuous ovens [79], utilization of the flue gases for preliminary heating of the material to be activated, its carbonization and the regeneration of spent adsorbents [14].

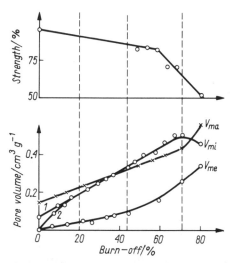

Fig. 2.6 Variation of the strength and pore volume of granules with burn-off: 1 — micropore volume with respect to benzene, 2 — micropore volume with respect to methanol (after Kostomarova and Surinova [62]).

In Figs. 2.6 and 2.7 graphs are presented illustrating the change of properties of spherical granules in the process of increasing levels of activation with steam in a rotary oven at 900 °C. We can distinguish four main regions of changes in the properties of the granules. When the burn-off is less than 20%, widening of the entrances to the pores main-

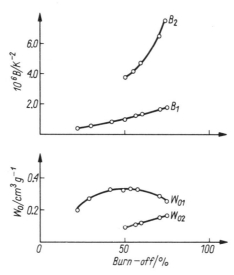

Fig. 2.7 Variation of the constants in the D–R equation with burn-off (after Kostomarova and Surinova [62]).

ly takes place, as confirmed by the equal volumes of micropores as determined by adsorption of benzene and methanol vapours. For burn-offs of 20–45% we observe a uniform development of micro- and mesopores as manifested by the gradual increase of the B constant of the D–R equation. For burn-offs of 45–70%, supermicropores appear, the volume of mesopores increases rapidly, and the strength of granules decreases. At burn-offs exceeding 70% the increase of the pore volume is due solely to the increase of meso- and macropores, and a further decrease of the granule strength is observed.

The high mechanical strength of spherical carbonized granules enables their activation in a fluidized bed. On activation under such conditions, active carbons of different porous structure were obtained; viz. (i) mechanically resistant materials with a large micropore volume and limited mesopore volume, (ii) materials with developed micro- and mesoporosity, and (iii) materials with large macropore volumes.

The modernization of adsorbent activation technology has as its aim increase of efficiency at all stages of the process. However, it also seeks the introduction of processes that contribute more effectively to the well-developed and mechanically-resistant porous structure of active carbons [80]. Understanding of the effect of the type of raw material used as well as of the conditions of carbonization and activation on the properties of the final product is paramount. Dziubalski, Korta and

Laśon describe [81] an experimental laboratory set-up for investigating the pyrolysis and activation processes (see Fig. 2.8). The main part of the set-up consists of two vertical steel pipes connected by a horizontal channel through which the basket with the sample can be transported from one pipe to the other. One of the pipes, which is the reactor, is placed in an electric oven, the second one, which serves as the cooling chamber, is all the time filled with nitrogen. The temperature in the reactor is determined by a thermocouple connected to the temperature-control device. Into this pipe a suitable gas is introduced at a controlled rate. In the case of pyrolysis, nitrogen is used, and in the case of activation, steam, carbon monoxide or their mixtures with nitrogen.

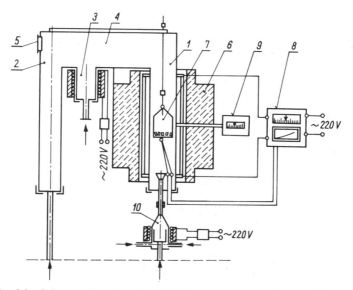

Fig. 2.8 Scheme of an experimental laboratory set-up for studying the pyrolysis and activation process: 1 — reactor, 2 — cooling chamber, 3 — sample heater, 4 — connecting channel, 5 — door for introducing the sample, 6 — electric oven, 7 — basket with the sample, 8 — device for control and measuring temperature inside the reactor, 9 — oven temperature control, 10 — gas mixing chamber (after Dziubalski, Korta and Laśon [81]).

The main advantage of this set-up is that the basket with the sample can be rapidly transported from the reaction pipe to the cooling pipe filled with nitrogen and kept at room temperature. This enables interruption of the pyrolysis or activation process conducted in the reaction pipe immediately after the pre-determined time has elapsed and 'freezing' of the carbon structure obtained.

The authors claim that their set-up makes it possible to determine the dependence of the carbonizate burn-off process, taking place in the course of activation, on:

(i) the kind of raw material used for obtaining the carbonizate,
(ii) conditions of pyrolysis,
(iii) volatile matter and ash content in the carbonizate,
(iv) activation temperature,
(v) thickness of the carbonizate layer (in the case of a monoparticle carbonizate layer, its activation conducted in the set-up simulates the fluidized bed activation process),
(vi) type of activating gas and its composition,
(vii) flow rate of the gas through the reaction pipe.

Another important activation process is chemical activation, where non-carbonized raw materials such as peat and sawdust are used. Waste carbon precipitate from decolorizing processes is also suitable. The conversion of such a raw material into active carbon proceeds under the action of dehydrating reagents at high temperatures. Then oxygen and hydrogen are selectively and almost completely removed from the carbon material, carbonization and activation proceeding simultaneously (usually at temperatures below 650°C).

The materials carbonized in this way are characterized by a reduced oxygen and hydrogen content and therefore their activation by means of inorganic materials is much more difficult as compared with that of non-carbonized materials. Wood — a material suitable for chemical activation — contains about 43% of oxygen and about 6% of hydrogen calculated per dry mass of the demineralized product. Brown coals contain 25% and 5% of oxygen and hydrogen, respectively. Industrially, phosphoric acid, zinc chloride, and potassium sulphide are mainly used as activating reagents [57]. Besides, other reagents exhibiting dehydrating activity may be used such as potassium thiocyanate and sulphuric acid, and other substances, which have not as yet found wider commercial application, such as metallic sodium or potassium, potassium carbonate, calcium oxide, calcium hydroxide, ammonia, ammonium chloride, aluminium chloride, iron and nickel salts, sulphur, chlorine, hydrogen chloride, hydrogen bromide, nitric acid, nitrogen oxides (sometimes together with sulphur dioxide), phosphorus pentoxide, arsenic pentoxide, borates, orthoboric acid and potassium permanganate. These substances often also reveal catalytic activity.

Activation with phosphoric acid is conducted as follows: the finely ground carbonaceous material is mixed with a phosphoric acid solution, the mixture is dried and heated in a rotary oven at 400–600°C; processes are, however, described where the heating temperature is higher (up to 1100°C). To obtain widely-porous active carbons used for decolorizing solutions, more phosphoric acid should be added than for carbons used in purifying gases or potable water. If zinc chloride is used as the activating agent, 0.4–5 parts of $ZnCl_2$ in the form of concentrated solution are mixed with 1 part of the carbonaceous material and the mixture is heated to 600–700°C (a flowsheet of the process is shown in Fig. 2.9). The use of zinc chloride in the activation process has been limited recently in view of environmental requirements.

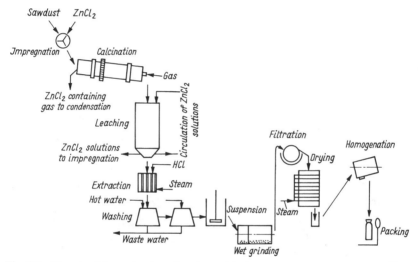

Fig. 2.9 Flowsheet for activation of carbon with zinc chloride (after Smišek and Černý [57]).

In the activation of carbons used in medicine, potassium sulphide and mixtures of potassium hydroxide and sulphur are used. The process is conducted in the absence of air in rotary ovens at 800–900°C. After extraction of potassium sulphides, treatment with dilute hydrochloric acid and washing to remove the chloride anions, the carbons are exposed to thermal treatment at 500–600°C in order to remove the sulphur. It should be noted that the complete removal of the activating agents from the activated carbon presents significant difficulties apart from which chemically-activated carbons often have a comparatively high ash content. Another disadvantage of this method for obtaining active carbons are the difficulties involved in recovering the chemical reagents which increases the overall manufacturing costs.

Among the advantages of chemical activation are: the relatively short time of activation, the high yield of active carbon and the good adsorption properties of the final product. However, depending on the type of activator, the results of chemical activation may differ considerably. For instance activated carbons obtained from sawdust by addition of potassium carbonate and heating at 800°C feature high adsorption of methylene blue and a large iodine number but low decolorizing capacity. Carbons activated with zinc chloride reveal good decolorizing properties but their methylene blue adsorption and iodine number are low [1].

2.4 ESTIMATION OF THE PROPERTIES OF ACTIVE CARBONS

The commercial use of active carbons, their transport, storage and sales require knowledge of the properties of these materials. The methods for estimating these properties are approved by the members of the Activated Carbons Sector Group of the European Council of Chemical Manufacturers' Federations (CEFIC). The tests require highly professional laboratories and advanced equipment. Most of the testing methods have been developed and approved by such organizations as the American Society for Testing Materials (ASTM), the American Water Works Association (AWWA), the Deutsches Institut für Normung e.V. (DIN), or the International Organization for Standardization (ISO).

2.4.1 Collection and Preparation of Samples

The reliability of the results of analytical tests depends largely on accurate adherence to the sampling procedure. In the cases of granular active carbons it is of paramount importance to determine not only the particle size distribution but also the differences in adsorptivity and density of the particular fractions. If during packing or transport in containers, the carbon separates into particle size fractions, then known statistical methods of collecting samples should be applied. For sampling of granular active carbons collected from individual containers or adsorption layers in operation, simple samplers are used. If the samples collected exceed the amount of product necessary for the particular tests, as when samples are collected directly from the stream of the manufactured product, they may be reduced by applying a mechanical sample reduction device.

The errors made in the stage of sampling or sample reduction cannot be compensated for by the accuracy of the particular tests. We should always be aware of this.

The methods of determining the principal properties of active carbons involve three main groups of tests: physical, adsorption, and chemical and physicochemical. The methods presented here are not exhaustive, even of those used in practice, and detail only the most important ones. Some more sophisticated and complex methods of describing the properties of active carbons are presented in later chapters.

2.4.2 Physical Tests

Bulk density. The bulk density is defined as the mass per unit volume of the active carbon sample in air including both the pore system and the voids between the particles. It is expressed in kg m^{-3} per dry mass. The bulk density of active carbon, depending on the shapes, sizes and densities of the individual particles, is indispensable for determining the size of unit packages. Two different standard methods are used for preparing a unit volume of powdered or granular active carbon. In the case of the powdered carbons the DIN ISO 787 method is applied [82], and for granulated carbons the ASTM D 2854 procedure is used [83].

Absolute density. The absolute or helium density is defined as the mass per unit volume of the carbon skeleton inaccessible to helium and is normally expressed in g cm^{-3}.

The volume of the carbon skeleton of the sample of known mass is determined from the volume of displaced gas (helium). The details of this test are discussed in Chapter 8. The absolute densities of commercial active carbons lie in the range of 2.0–2.1 g cm^{-3}, while that of graphite is 2.26 g cm^{-3} which value has not been attained for active carbons.

Particle density. Particle density, also known as mercury density, is defined as the mass per unit volume of carbon particles including the pore system and is expressed in g cm^{-3}. Mercury is used for measuring this quantity since it completely fills the spaces between the particles of sample. The measurement is carried out without applying excess pressure. The small error due to the possible filling of some macropores with mercury is neglected. The values of the particle density obtained in this way for active carbons lie in the range 0.6–0.8 g cm^{-3}, while that for cokes is about 0.9 g cm^{-3}. The measurements of particle density are carried out in terms of simple pyknometry of solids with the use of mercury as the medium. For details see Chapter 8.

Particle size. Granular carbons. For particle size classification, use is made of screen analysis usually carried out on screens put in a swinging motion for a given time. The additional impacts may affect somewhat the results of analysis due to the crumbling of the carbon particles. Screens with standard mesh dimensions are used (DIN 4188) [84]. The particle size distribution of the sample is given in weight per cent. A ±5% deviation from the specified particle size is usually acceptable.

Extruded carbons. The fractionation method used for crushed carbons can be fully applied to fractionating cylindrical granules. It may happen, however, that these granules, whose length is greater than their diameter, will assume a vertical position in the course of screening and pass through the screen. This takes place especially then when the layer of the carbon material on the screen is very thin. In such a case the result of screen analysis is influenced by the ratio of the granule diameter to length, and the analysis does not give accurate information about the granule length distribution. Therefore this procedure allows us only to conclude about the presence of fine particles (subparticles) in the sample. Hence screen analysis, e.g. by the ASTM D 2682 method [85], may be used successfully in all cases where comparative data are necessary, as in routine production control.

For accurate determination of granule length and diameter distribution, a typical sample is photographed, and the dimentions of the particles are calculated by means of an automatic computing device. However, this method is seldom used in conventional control.

Powdered carbons. Since it is difficult to fractionate powdered carbons with a particle size below *ca.* 0.1 mm on screens, wet screening is usually applied. For this purpose an aqueous suspension of the powdered carbon sample is boiled for some time in a beaker to wet the surface of the particles after which it is passed through a set of microscreens with fine meshes pre-wetted with water. The fractionation is carried out by washing with a deionizing agent. After drying the particular fractions are analyzed gravimetrically.

Air separators may also be used for determining the particle size distribution of powdered carbons, however, in such a case the result may be shifted towards greater sizes due to the formation of aggregates of wetted carbon particles which are not broken up in the air stream. Errors in the measurements may also be due to electrostatic charges.

Gas pressure drop over a packed carbon bed. The pressure drop gives information about the resistance to the flow of a gas through a carbon layer. The pressure drop over the carbon layer is determined by the modified semi-empirical Ergun equation which accounts for the shape and size of the particles, as well as the temperature, pressure and linear velocity of the flowing gas. The pressure drop is expressed in Pa m^{-1} of the carbon layer height. The modified Ergun formula used in the calculations has the form [86]:

$$\frac{\Delta p}{L} = K_1 \mu_T v_0 + K_2 \varrho_{T_p} v_0^2 \qquad (2.19)$$

where: Δp is the pressure drop (Pa), L — the length of the carbon column (m), v_0 — the linear velocity of air flow (m s^{-1}), μ_T — dynamic gas viscosity dependent on temperature T and pressure p (Pa s), ϱ_{T_p} — the gas density dependent on temperature T and pressure p (kg m^{-3}), T — gas temperature (K), p — gas pressure (Pa) $p \approx p_B + \frac{1}{2} \Delta p$, p_B — the barometric pressure (Pa). The values of L and v_0 are given, and those of μ_T and ϱ_{T_p} can be extracted from tables, K_1 and K_2 are constants found by measuring $\Delta \varrho$ at six different air flow velocities v_0 and calculated by the least squares method.

In most cases the resistance shown by the carbon bed to the passing water or air stream under normal conditions is information of practical importance. Such measurements are made with sufficient accuracy in the standard dynamic tube. To best reduce wall effects, the inner diameter of the tube should be at least 10 times greater than the diameter of the particles of the active carbon under test. The depth of the carbon bed should not be less than 20–30 cm, but usually is 50 cm. The degree of packing of the bed being important from the point of view of measurement accuracy, the densely packed beds are prepared by the vibrational or shaking method. The stream of air used for determining the resistance of the bed should firstly be purified from traces of oil from the compressor and dried. To prevent the stream of gas or liquid from disturbing the ordering of the carbon bed, especially when high stream velocities are applied, the fluid is fed downstream.

Mechanical strength. The manufacturers and users of active carbons use methods for determining the mechanical strength of carbon that often differ significantly from one another. The majority of methods of testing the mechanical strength of active carbons (AWWA [87], ICUMSA [88], ASTM D 3802 [89]) are based on the following principle: the sample of active carbon is subjected to a mechanical load, and the strength is estimated from the amount of fine fractions of carbon generated or the decrease of the mean size of the carbon particles. Among the most common tests used are:

— Ball-mill hardness: the carbon is abraded for a given time in a horizontal cylinder with steel or ceramic balls under prescribed conditions.

— Abrasion strength: the carbon is abraded by an iron rod in a horizontal rotating cylindrical sieve of given dimensions for a prescribed time.

— Impact hardness: carbon particles are broken by dropping a weight onto a sample under controlled conditions.

— Ball-pan hardness: the carbon is shaken for a given time in a pan together with a number of steel balls of known diameter.

— T-bar hardness: the carbon is abraded in a cylindrical container with a T-shaped stirrer with a prescribed rotation speed and for a prescribed time.

— Crushing strength: the pressure applied to crush a granule of carbon (not applicable for broken granules); commercially available apparatus is used.

— Impact hardness (fluidized bed): the carbon is pneumatically agitated for a standard time in a vertical cylinder, the top of which is provided with an impact plate.

2.4.3 Adsorption Tests

The adsorption properties of active carbons are generally estimated by determining the isotherms of adsorption from the liquid phase. The determination of the adsorption of one test substance from an aqueous solution is often insufficient for characterizing the adsorption properties of a carbon. Thus the properties of active carbons are estimated by comparing the results of measurements for different adsorbates, e.g. by comparing the adsorptions of fairly large molecules of methylene blue or iodine.

Phenol adsorption isotherm. This isotherm provides information on the adsorptive properties of active carbon, e.g. in water treatment. A detailed description of the determination is given in ASTM 1783-70 [90] and in DIN 19603 [91]. According to the DIN procedure the adsorption isotherm described by the experimental Freundlich equation is determined for different weighed-out portions of the powdered carbon. Usually the following form of this equation is used:

$$\frac{a}{M} = K c_P^{1/n} \qquad (2.20)$$

where: a is the quantity of phenol adsorbed, M — the weight of active carbon, c_P — the residual phenol concentration in the solution after adsorption (the concentration of phenol in the solution changes from the initial one c_0 to the final one c_P), K and $1/n$ are constants for a given adsorption system. Next, the adsorption capacity of the carbon is found, graphically or analytically (from the linear form of Freundlich's equation):

$$\log \frac{a}{M} = \log K + \frac{1}{n} \log c_P \qquad (2.21)$$

for the equilibrium concentration of phenol in a solution of 1 mg dm^{-3}. This value is assumed to be equal to the phenol adsorption capacity of active carbon. Usually, to allow the estimation of the isotherm slope, the adsorption capacity at a phenol concentration of 0.1 mg dm^{-3} is also given.

Often the phenol adsorption capacity of the active carbon is determined in wt.% of the adsorbed phenol from the equation

$$Q_p = (c_0 - c_P) \frac{V}{10 \, M} \qquad (2.22)$$

where: Q_p is the quantity of phenol adsorbed by the carbon in wt.%, V — the volume of the phenol solution in dm^3, M — the weight of the active carbon in g, c_0 and c_P are the initial and final (residual) phenol concentrations in the solution, respectively.

If a full characterisation of active carbon used for water treatment is required, then knowledge of the adsorption capacity for phenol concentrations from 1 to 0.1 mg dm^{-3} is insufficient, and derivation of the so-called phenol number by the method given by AWWA [92] should be undertaken. According to this procedure the quantity of active carbon (in mg) necessary for lowering the phenol concentration in aqueous solution from 100 to 10 mg is determined. These calculations are also made with the use of the Freundlich isotherm.

Iodine adsorption. The study of the process of iodine adsorption and also the determination of the iodine number is a simple and quick test for estimating the specific surface area of active carbon. The iodine number is defined as the number of milligrams of iodine adsorbed by 1 g of active carbon from an aqueous solution when the iodine concentration of the residual filtrate is 0.02 N. If the final values obtained are different from

0.02 N but lie in the range of 0.007–0.03 N, appropriate corrections are necessary. In this method it is assumed that iodine at the equilibrium concentration of 0.02 N is adsorbed on the carbon in the form of a monolayer, and this is the reason why there is a relationship between the iodine number of active carbon and its specific surface area which may be determined, for example, by the BET method.

The specific surface areas of active carbons with highly developed microporous structures as determined by the iodine number method are too low. This is because iodine is adsorbed chiefly on the surface of pores much larger than 1 nm, while in active carbons with large specific surface areas the proportion of very fine pores inaccessible to iodine molecules is significant. One of the methods of determining the iodine number that can be utilized is the method described in the AWWA standard B 600–78 [92]. In this method the iodine number I_n of active carbon is calculated from the formula:

$$I_n = \frac{a}{M} A \qquad (2.23)$$

where: a is the quantity of iodine adsorbed by the carbon in mg; $a = 12.693\, N_1 - 279.246\, N_2 V$, where N_1 is the normality of the iodine solution, N_2 is the normality of the sodium thiosulphate solution, V is the volume of the thiosulphate solution in cm^3; M is the mass of active carbon in g, and A is a correction factor depending on the residual normality of the iodine solution other than 0.02 and is to be found in tables.

Methylene blue adsorption. The methylene blue value gives an indication of the adsorption capacity of an active carbon for molecules having similar dimensions to methylene blue, it also gives an indication of the specific surface area of the carbon which results from the existence of pores of dimensions greater than 1.5 nm. The molecule of methylene blue

$$\left[(CH_3)_2 N - \underset{S}{\underset{|}{\bigodot}} - N(CH_3)_2 \right]^+ Cl^-$$

has linear dimensions greater than 1.5 nm, but adsorption experiments on silica gels with a laminar structure of the lattice have shown that the molecule of this dye is adsorbed in pores of exactly these dimensions, as if it were a flat plate [1].

The methylene blue value is defined as the number of cubic centimetres of standard methylene blue solution decolorized by 0.2 g of the active carbon (dry basis). The standard solution is prepared by dissolving 1.5 g of methylene blue (preferably of Merck medical, chemically pure, free from zinc chloride grade) so as to obtain 1 dm^3 of solution. According to the DAB 7 method [93], the standard methylene blue solution is added portionwise over 5 min with shaking to 200 mg of active carbon ground (< 0.1 nm) and dried at 150°C. The methylene blue value is the number of cubic centimetres of the solution that have been decolorized.

In the USA the methylene blue value is determined by titrating 15 mg of powdered active carbon with the methylene blue solution (1 g dm^{-3}) until the solution ceases to decolorize after 5 min.

In the USSR adsorption of methylene blue is conducted using 1 g of active carbon and a 0.15% solution of the dye.

In Japan the standard method is based on adsorption of methylene blue from a solution of concentration of 1.5 g dm^{-3}.

Phenazone adsorption. This is a test to estimate the adsorption capacity of activated carbon for pharmaceutical purposes. Phenazone (2,3-dimethyl-1-phenyl-5-pyrazolone) has the structure

Phenazone adsorption is defined as the mass of phenazone (in g) adsorbed during 15 min by 100 g of dry active carbon from an aqueous solution [94].

The test is conducted by shaking for 15 min 0.30 g of powdered (< 0.1 nm) and dried (150°C) active carbon with 25 cm^3 of a 1% aqueous solution of phenazone. Then 0.5 g of potassium bromide and 20 cm^3 of dilute hydrochloric acid are added to 10 cm^3 of the filtrate. Titration is made with a 0.1 N solution of potassium bromide/bromate using 0.1 cm^3 of ethoxychrysoidine as indicator until the colour changes from raspberry red to pinkish yellow. The titrating solution should be added very slowly (1 drop every 15 sec). A blank test should be carried out in parallel with 10 cm^3 of phenazone solution.

Milligram value. This is a test to estimate the adsorption capacity of activated carbon for decolorizing purposes. The milligram value (mV) is defined as the number of milligrams of the active carbon under test necessary to achieve a degree of decoloration of 200 cm³ of a molasses solution analogous to that achieved with a standard carbon.

2.4.4 Chemical and Physico-Chemical Tests

Volatile matter content. The international standard used for determination of volatile matter in hard coal and coke is also applicable to active carbon [95]. A sample of powdered (< 0.1 nm) active carbon is heated at 900°C for 7 min. The percentage of volatile matter x_d (in wt. %) is determined from the equation:

$$x_d = 100 \frac{100(B-F) - M_c(B-G)}{(B-G)(100-M_c)} \qquad (2.24)$$

where: B is the mass in g of the crucible, lid and carbon sample before heating (1.000 ± 0.010 g), F is the mass in g of the crucible, lid and carbon sample after heating, G is the mass in g of the empty crucible and lid, M_c is the moisture, as a percentage by mass, in the sample, determined according to the method specified below.

Moisture content. The oven drying method is used when water is the only volatile material present in the active carbon. The xylene distillation method or Karl Fischer method is used when the carbon is known or suspected to be heat-sensitive or to contain volatile matter other than water.

A simple method of determining the water content is drying active carbon in a dryer (ASTM D 2867-70) [96]. The sample of powdered (1–2 g) or granular (5–10 g) carbon is dried at 150°C to constant weight (usually about 3 h). The water content in percent is calculated from the equation:

$$M_c = 100 \frac{B-F}{B-G} \qquad (2.25)$$

where: B is the mass in g of container with lid plus original sample, F is the mass in g of the container with lid plus the dried sample, G is the mass in g of the container and lid.

Ash content. The ash content of active carbon can be determined by ignition of the crucible in an electric muffle furnace (ASTM D 2866-70) [97]. Ignition is conducted at 650 ± 25°C for 3 to 16 h, depending on the type of active carbon and the dimensions of its particles, to constant mass. The ash content A_c in wt. % is found from the equation:

$$A_c = 100 \frac{F-G}{B-G} \tag{2.26}$$

where: G is the mass of the empty crucible in g, B is the mass of the crucible plus the dried sample in g, F is the mass of the crucible plus the ashed sample in g.

pH value. Active carbon carrying inorganic matter and chemically active oxygen groups on its surface may alter the pH of liquid systems to which it is added. Since active carbons finding application in the chemical and food industries are used to purify various substances sensitive to changes of pH, this quantity is often the crucial parameter of the active carbon when selecting it for a given purpose.

The test consists in weighing out a 4 g portion of active carbon into a beaker, adding 100 cm^3 of distilled, CO_2-free water, and boiling for 5 min. The contents are left to stand until it cools down a little and the carbon settles. The supernatant liquid is poured off and its pH is measured with a pH-meter with a glass electrode.

2.5 CHARACTERISTICS AND APPLICATIONS OF COMMERCIAL ACTIVE CARBONS

Data on the world production of active carbons are given in Table 2.5 [3]. It is obvious that the output and the types of carbon manufactured depend strictly on the existing demand. In this section we discuss, by way of example, the main grades of active carbons manufactured by the Dutch Company Norit for specific industrial applications. We chose this company first of all because it is recognized in Europe as the manufacturer of the widest range of active carbons on a large commercial scale, and secondly because these carbons exhibit very good properties. Of course our choice is entirely arbitrary and somebody could try to present the spectrum of applications of active carbons using as an example of the product range of some other manufacturer, however, this would be equally arbitrary.

Several other manufacturers of active carbons are mentioned in Table 2.1. It is our intention to show the reader how wide and versatile are the applications of particular grades of active carbon, and how variable can be the adsorption and sometimes also the catalytic properties of these carbons.

Table 2.5
World production of active carbon (according to 1974 data) (after Jüntgen [3])

Region	Output, kilotonnes/year			Comments
	total	powdered	granular	
North America	127	75	52	7 plants, however, 3 plants account for 87% of production
Japan	40	25	15	20 plants; in the period 1968–1973 the production doubled
Asia (without Japan) and South America	20			
Western Europe	100	75	25	10 plants, however, 4 plants account for 85% of production
Eastern Europe and USSR	lack of data			Carbon is manufactured in Czechoslovakia, Poland, Yugoslavia, Hungary and the USSR

For water treatment, encompassing dechlorination, deoiling of condensate, production of potable water, preliminary treatment prior to ion exchange on synthetic resins, purification on sand beds, purification of industrial waste waters, Norit recommends the following main grades of active carbon [98]:
— extruded carbon: ROW 0.8 Supra,
— crushed carbons: RK 1–3, PK 0.25–1 and PX 3–5,
— powdered carbons: AZO, W20 and W52.

Norit ROW 0.8 Supra is a peat-based granulated active carbon which can be regenerated (reactivated) after adsorption. In view of its large total volume of the pores and suitable distribution of the pore volumes as a function of their linear dimensions, this carbon is particularly suitable for improving the taste of potable water, removing the disagreeable smell of chlorine, ozone, micropollutants, and dissolved organic substances. This active carbon has the following characteristic:

		Refs.
Apparent density/g dm^{-3}	380	[83]
Density backwashed drained/g dm^{-3}	335	
Moisture/%	2	[96]

Ash content/%	6	[97]
Phenol adsorption/%	6	[91]
Iodine number/mg g^{-1}	1050	[92]
Total pore volume/cm^3 g^{-1}	1.0	

In the food industry, active carbon is applied in the production of agar, alcohol and alcoholic beverages, beer, cocoa butter, caffeine, gelatin, glucose, lactose, fruit juices, pectins, refined sugar, vegetable oils and fats, wine yeast, etc. For these purposes Norit recommends the following main grades of active carbons with very good adsorptive, mechanical and other properties [98]:

— extruded carbons: ROW 0.8, ROX 0.8, RAX 1,
— crushed carbons: PK 0.25–1, PK 1–3,
— powdered carbons: SX Plus, SX2, SA2, SA4, D10, ZN2, ZN5, PN2, CA1/CN1, Glucoblend 1/2, CA3/CN3.

In this group the powdered carbon Norit SA4 deserves particular attention. It is a steam-activated carbon that can be used in a wide range of applications. This carbon is especially suitable for potable water treatment, decolorization and purification of vegetable oils and fats, inorganic acids and salts, organic acids, alkaline electroplating baths, etc., and for improvement of the smell and taste of beverages such as wine, beer, fruit juices. Norit SA4 has the following characteristics:

		Refs.
Apparent density/g dm^{-3}	490	[99]
Moisture/%	2	[96]
Ash content/%	6	[97]
Phenol adsorption/%	4	[91]
Molasses number	525	
Methylene blue adsorption/%	11	
Iodine number/mg dm^{-1}	750	[100]
Total internal surface area (BET)/m^2 g^{-1}	650	
pH	alkaline	
Particle size (wet sieving):	%	
> 10 μm	80	
> 44 μm	37	
> 74 μm	20	
> 150 μm	5	

In the pharmaceutical industry, the main applications of active carbon are the manufacture of antibiotics and sulphonamides, haemoperfusion, production of injection fluids, pharmaceutical

intermediates, vitamins B and C, etc. Norit recommends for these applications the following grades:
— extruded: ROX 0.8, RBXS1,
— powdered: SA Plus, SX Plus, SX2, CA1/CN1, CA3/CN3.

One of the active carbons used in the pharmaceutical industry is the steam-activated Norit SA Plus with high adsorption capacity owing to which it has a wide range of applications. It is especially useful for decolorizing concentrated products of the chemical industry such as inorganic acids (e.g. phosphoric acid) and enzymes. It is also used for improving the quality of alcohols, as well as purifying and decolorizing products of the pharmaceutical industry. Norit SA Plus has the following characteristics according to the manufacturer's tests [98]:

Apparent density/g dm^{-3}	340
Moisture/%	2
Molasses number	225
Methylene blue adsorption/mg g^{-1}	210
Iodine adsorption/mg g^{-1}	1000
Total internal surface area (BET)/m^2 g^{-1}	1000
Ash content/%	10
pH	alkaline
Particle size:	%
> 10 μm	73
> 44 μm	26
> 74 μm	10
> 150 μm	2

It appears that active carbon finds its largest applications in the chemical industry in the production of inorganic acids (e.g. hydrochloric and phosphoric), organic acids (e.g. acetic, citric, fumaric, glutamic, lactic, nicotinic, tartaric), acrylic amides, absorption fluids, dry cell batteries, caprolactam, solvents for dry cell batteries, dyestuffs (e.g. azo compounds), electrolytic baths, ethanolamines (MEA, DEA, etc.), fatty acids, glycerin, glycols, polyamide fibres, melamine, pentaerythritol, phosphates, phthalates (DOP, DNP, etc.), and waxes.

Active carbons are also used for recovering gold, reclamation of lubricants, and the absorption of mercury vapour from air.

The following Norit products are recommended for use in the above-mentioned fields of chemical production [98]:
— extruded carbons: RO 0.8, ROX 0.8, R 2020,
— crushed carbons: PK 3–5, PK 1–3, PK 0.25–1, Elorit,

— powdered carbons: AZO, SX 2, SA Plus, SA 2, CA1/CN1, CA3/CN3.

The active carbons Norit AZO and Elorit feature very interesting properties. Norit AZO is a steam-activated carbon with a coarse particle structure. It is used in processes where good filtering properties are important, especially where normal filter aids fail, e.g. in the removal of mercury from caustic soda solutions in the mercury electrolysis process. The main properties of Norit AZO are:

			Refs.
Bulk density/g dm^{-3}		310	[83]
Moisture/%		2	[96]
Total internal surface area (BET)/m^2 g^{-1}		700	
Ash content/%		14	[97]
Particle size:	%		
> 44 μm	99		
> 74 μm	78		
> 150 μm	37		
> 250 μm	5		

Elorit is a peat-based steam-activated granular carbon especially suitable for the manufacture of dry cell batteries. This carbon has the following characteristics:

			Ref.
Bulk density/g dm^{-3}		280	[83]
Moisture/%		2	[93]
Total internal surface area (BET)/m^2 g^{-1}		700	
Ash content/%		14	[97]
Ca content/%		0.5	
Mg content/%		0.5	
Fe content/%		0.3	
pH		alkaline	
Particle size:	%		
> 0.50 mm	0.5		
> 0.25 mm	45		
> 0.105 mm	92		

Active carbons are widely used in the recovery of solvents and in the treatment of air in air conditioning plant, especially for offices, computer rooms, cold stores. They provide the main element of gas masks for military and industrial use. Among the latter we distinguish gas masks of type A, B, E and K absorbing organic, inorganic, acidic and ammonia vapours, respectively. Active carbons are also useful for removing hydrocarbons, kitchen vapours (in ventilating hoods), controlling odours and removing ozone. They are also used to recover solvents, styrene and vinyl chloride monomers, and to absorb fermentation and warfare gases.

The following Norit products are recommended for the above purposes [98]:
— extruded carbons: RO, R, RB, R2030, R Extra, Sorbonit, Sorbonorit B,
— extruded impregnated carbons: RB AA, RFZ, RGM, RZN,
— binding agent: Norithene.

As an example of an active carbon suitable for the above applications we may mention Sorbonorit 3 which is a steam-activated extruded carbon in the form of 2.9 mm pellets. All the active carbons of Sorbonorit type (2, 3, and 4) are characterized by the high strength of the pellets. The favourable adsorption and desorption properties of these carbons make them particularly suitable for solvent recovery operations, and also for other gas-phase applications involving regeneration. Regeneration can be done by steam, with an inert gas or desorption via pressure-swing techniques.

The characteristics of Sorbonorit 3 are as follows:

		Refs.
Apparent density/g dm^{-3}	400	[83]
Pellet diameter/mm	2.9	
Moisture/%	2	[96]
Ash content/%	5	[97]
Abrasion resistance/%	99	[89]
Pore size distribution		
micropores (< 1 nm)/cm^3 g^{-1}	0.47	
mesopores (1–100 nm)/cm^3 g^{-1}	0.10	
macropores (> 100 nm)/cm^3 g^{-1}	0.47	
Total internal volume area (BET) from		
benzene adsorption/m^2 g^{-1}	1200–1300	
Ignition point/°C	450	[101]

An important application of active carbon is gas purification processes. There are used to remove acid, ammonia, propane and butane vapours, carbon dioxide, chlorine, cyanogen vapours, hydrocarbons, phosgene, radioactive iodides, oil from compressed gases, and to

separate gases, e.g. xenon from krypton. Norit recommends for the above purposes the following grades [98]:
- extruded carbons: R, RB, R 2030, R Extra, RO,
- extruded impregnated carbons: RBAA, RGM, RKJ, RZN, RFZ,
- binding agent: Norithene.

Among these grades Norit RB 3, which is an extruded, steam-activated carbon, is particularly recommended to industry for removal of low concentrations of contaminants from gases, and in those cases where a thick carbon layer is required. As examples of the application of this grade of carbon we can cite: the purification of process gases from exhaust gas, removal of carbon dioxide from cold stores, removal of oil from air or gas, and removal of hydrogen sulphide from sewage air.

The characteristics of Norit RB 3 steam activated carbon are as follows:

		Refs.
Apparent density/g dm^{-3}	460	[83]
Pellet diameter/mm	2.9	
Moisture/%	2	[96]
Ash content/%	6	[97]
Abrasion resistance/%	99	[89]
Pore size distribution		
micropores ($<$ 1 nm)/cm^3 g^{-1}	0.36	
mesopores (1–100 nm)/cm^3 g^{-1}	0.08	
macropores ($>$ 100 nm)/cm^3 g^{-1}	0.41	
Total internal surface area (BET) from		
benzene adsorption/m^2 g^{-1}	1000	
Ignition point/°C	$>$ 470	

Finally we should like to mention the application of active carbons as catalysts or catalyst supports for example in hydrogenation processes where noble metals are deposited on the carbon, e.g. in vinyl chloride and vinyl acetate production. Active carbon is also used as a catalyst in the removal of mercury or sulphur.

Norit recommends the following main types of active carbon for use in chemical catalysis [98]:
- extruded carbons: RO, R Extra, RKD, ROX,
- extruded and impregnated carbons: RBS, RFN, RFZ, RKD with zinc acetate,

— crushed carbon: PKDA,
— powdered carbon: SX.

As an example of a carbon suitable for catalysis we can mention Norit R1 Extra which is a steam-activated extruded carbon. The appropriate grain size and adequate pore volume distribution as a function of their linear dimensions make this carbon especially suitable for use in catalysis and in gas masks. The large volumes of the particular kinds of pores make it amenable to impregnation. Nort R1 Extra has the following characteristic parameters:

		Refs.
Apparent density/g dm^{-3}	435	[83]
Pellet diameter/mm	1	
Moisture/%	2	[96]
Ash content/%	8	[97]
Abrasion resistance/%	97	[89]
Pore size distribution:		
micropores (< 1 nm)/cm^3 g^{-1}	0.47	
mesopores (1–100 nm)/cm^3 g^{-1}	0.10	
macropores (> 100 nm)/cm^3 g^{-1}	0.47	
Total internal surface area (BET) from		
benzene adsorption/m^2 g^{-1}	1200–1300	
Ignition point/°C	450	[101]

In Poland there are two plants manufacturing active carbons, namely the Destructive Wood Distillation Plant and the ZEW Carbon Electrodes Plant. The former manufactures various types and grades of extruded active carbons, from which many carbon sorbents are also produced. The annual productive capacity of this plant amounts to about 1000 t [80]. The production range encompasses four main grades of active carbons:

(i) Carbon A with pellet diameter of 1.7 mm,
(ii) Carbon AHD with pellet diameter of 1.7 mm and high adsorptivity with respect to organic substances,
(iii) Carbon AG-5 with pellet diameter of 1 mm, and
(iv) Carbon N with pellet diameter of 3.5 mm.

Grade A is an extruded active carbon obtained from powdered coal and hardwood tar by pressing under elevated pressure, followed by drying, carbonization and steam activation at high temperature. This grade of carbon has the following parameters:

Particle size distribution — remainder on screen with mesh diameter:

2.75 mm/%	max. 8
1.50 mm/%	not limited
1.00 mm/%	max. 8
< 1.00 mm/%	max. 0.6
Apparent density/g dm^{-3}	500
Abrasion resistance/%	min. 90
Water absorption/cm^3 g^{-1}	min. 0.8
Protective activity towards benzene vapour	min. 45

Grade A carbon is used as adsorbent of organic vapours, as well as a carrier of various sorbents, e.g. for absorption of acid gases, hydrogen sulphide, ammonia, and sulphur dioxide.

Grade N is an active carbon obtained from a special type of hard coal which is ground and extruded with addition of a binding agent. The granules are subjected to carbonization and activation in rotary ovens. This grade of carbon has the following characteristics:

Particle size distribution — remainder on screen with mesh diameter:

6.3 mm/%	0
3.6 mm/%	min. 90
2.8 mm/%	min. 7
1.0 mm/%	max. 2
< 1.0 mm/%	1
Apparent density/g dm^{-3}	430
Abrasion resistance/%	95
Water absorption/cm^3 g^{-1}	min. 0.8
Ash content/%	*ca.* 18
Static adsorption capacity/%	min. 25

Grade N carbon is used for recovering readily flammable solvents such as: petrol, toluene, alcohols, ethers, and trichloroethane. It finds application in removing undesirable and toxic components (e.g. sulphur, ammonia) from industrial gases, as a catalyst support in contact reactions, and in the production of neutral gases by selective adsorption.

The ZEW plant manufactures powdered and crushed active carbons, mainly designed for adsorption from the liquid phase. The principal raw material used in this production is charcoal and to some extent semi-coke from hard coal. The annual output of powdered and crushed active carbons is *ca.* 6000 t.

Considering the methods of production, raw materials, and applications, the active carbons manufactured by the ZEW plant are classified as follows:

(i) Decolorizing carbons — Carbopols,
(ii) Carbons for water and sewage treatment,
(iii) Carbons for medical and pharmaceutical use,
(iv) Carbons for catalytic applications,
(v) Bone charcoal — Carboneks,
(vi) Briquetted active carbons for adsorption of vapours and gases,
(vii) Special grade active carbons — Akwaryt, Depolaryt, etc.

By way of example let us now consider in greater detail several grades of decolorizing carbons.

Carbopol H Extra is a carbon obtained from vegetable raw materials of good quality by multistep physico-chemical activation. The activated product is further improved by treatment with hydrochloric acid and water. This carbon has the following characteristics:

Milligram value (decolorizing properties)/mg	up to 170
Methylene blue adsorption/cm^3	35
Moisture/%	max. 15
Ash content/%	max. 3
HCl soluble matter/%	1
Maximum content of ions:	
$\quad$ Ca^{2+}/%	0.1
$\quad$ Fe^{3+}/%	0.08
$\quad$ SO$_4^{2-}$	none
$\quad$ NO$_3^-$	none
$\quad$ Cl$^-$/%	max. 0.015
pH — depending on requirement	3–6

Carbopol H Extra is used for decolorizing and purifying organic and inorganic chemical and pharmaceutical products.

Carbopol O Extra is an active carbon obtained from a raw material of vegetable origin of good quality, activated in a three-step physico-chemical process, and neutralized with gaseous sulphur dioxide. This powdered carbon has the following characteristics:

Milligram value/mg	up to 170
Methylene blue adsorption/cm^3	min. 33
Moisture/%	max. 15
Ash content/%	max. 8
pH	6–8

Carbopol O Extra has very good filtering properties and shows good wettability. It is used for decolorizing and removing the odour of vegetable oils and animal fats. It also finds application in purifying organic and inorganic salts.

Carbopol Z Extra is an active carbon obtained from a vegetable raw material of good quality. It is activated in a two-step physicochemical process. The finely powdered carbon of homogeneous particle size is a product of high quality and with a wide range of applications. This carbon has the following properties:

Milligram value/mg	up to 170
Methylene blue adsorption/cm^3	min. 35
Moisture/%	max. 8
Ash content/%	max. 7
pH	> 8

Carbopol Z Extra shows very good filtering properties. It is used for decolorizing and purification of foods, but also in the pharmaceutical and chemical industries for decolorizing, purification and adsorption.

References

[1] Kienle (von), H., Bäder, E., *Aktivnye ugli i ikh promyshlennoe primenenie* (Active carbons and their commercial application), Khimiya, Leningrad 1984.
[2] Kostomorova, M. A., Perederii, M. A., Surinova, S. I., *Khim. Tverd. Topl.* **2**, 5 (1976).
[3] Jüntgen, M., *Carbon* **15**, 273 (1977).
[4] Dubinin, M. M., Zaverina, E. D., *Zh. Prikl. Khim.* **1**, 113 (1961).
[4a] Choma, J., Jaroniec, M., Piotrowska, J., *Carbon* **26**, 1 (1988).
[4b] Matsumura, Y., Yamabe, K., Takahashi, H., *ibid.* **23**, 263 (1985).
[4c] Krebs, C., Smith, J. M., *ibid.* **23**, 223 (1985).
[4d] Dobrowolski, R., Jaroniec, M., Kosmulski, M., *ibid.* **24**, 15 (1986).
[4e] Alves, B. R., Clark, A. J., *ibid.* **24**, 287 (1986).
[4f] Voudrias, E. A. Larson, R. A., Snoeyink, V. L., *ibid.* **25**, 503 (1987).
[4g] Jaroniec, M., Choma, J., Świątkowski, A., *et al.*, *Chem. Engin. Sci.* **43**, 3151 (1988).
[4h] Carrot, P. J. M., Roberts, R. A., Sing, K. S. W., *Carbon* **25**, 59 (1987).
[4i] Dubinin, M. M., *ibid.* **25**, 593 (1987).
[4j] Roberge, P. R., Beaudoin, R., Berthiaume, J. F., *ibid.* **26**, 173 (1988).
[4k] Kaneko, K., Nakahigashi, Y., Nagata, K., *ibid.* **26**, 327 (1988).
[4l] Rodriguez-Reinoso, F., Martin-Martinez, J. M., Molina-Sabio, M., *et al.*, *ibid.* **23**, 19 (1985).
[4m] Martin-Martinez, J. M., Rodriguez-Reinoso, F., Molina-Sabio, M., *et al.*, *ibid.* **24**, 255 (1986).
[4n] Kraehenbuehl, F., Stoeckli, H. F., Addoun, A., *et al.*, *ibid.* **24**, 483 (1986).

[4o] Rozwadowski, M., Wojsz, R., *ibid.* **26**, 111 (1988).
[5] Onusajtis, B. A., *Obrazovanie i struktura kamennougol'nogo koksa* (Generation and the structure of coke), Izd. AN SSSR, Moscow 1960.
[6] Ettingier, J. L., Jankovskaya, M. F., *Uglerodnye adsorbenty i ikh primenenie v promyshlennosti* (Carbonaceous adsorbents and their application in industry), Izd. Tekhn. Instituta Lensovet, Perm 1969.
[7] Gęsior, G., *Proceedings of the VIth All Polish Adsorption Colloqium on Sorption and the Protection of Man and Environment,* Military Technical Academy, Warsaw 1981, p. 115.
[8] Chodyński, A., Turonek, M., *Sprzęt ochronny dróg oddechowych dla górnictwa i innych gałęzi przemysłu oraz technologie materiałów sorbujących* (Equipment for protecting the respiratory tracts is mining and other industries as well as the adsorption materials technologies), Proceedings GIG Katowice 1977.
[9] Roga, B., Tomków, K., *Chemiczna technologia węgla* (Chemical technology of coal), WNT, Warsaw 1971.
[10] Keltzev, A. V., Vorobev, T. I., Karolev, J. G., *et al., Koks i Khimiya* **8**, 29 (1976).
[11] Chodyński, A., Gąsior, G., *Koks, Smoła, Gaz* **23**, 315 (1978).
[12] Cadenhead, D. A., Everett, D. H., *Industrial Carbon and Graphite,* Pergamon Press, London 1958.
[13] Khotunstev, L. L., Berezkina, Z. A., Grebennikov, V. S., *et al., Khim. Tverd. Topl.* **2**, 58 (1967).
[14] Makhorin, K. J., Tishchenko, A. T., *Vysokotemperaturnye ustanovki s kipyashchim sloem* (High temperature fluidized bed furnaces), Tekhnika, Moscow 1966.
[15] Matcalfe, J. E., Kawahata, M., Walker, P. L. Jr., *Fuel* **42**, 233 (1963).
[16] Patel, R. L., Nandi, S. P., Walker, P. L. Jr., *Fuel* **51**, 47 (1972).
[17] USA Patent 2894914, 1959.
[18] Walker P. L., Austin, L. G., Nandi, S. P., *Chem. Phys. Carbon* **2**, 257 (1966).
[19] Singh, D. D., *Chem. Ind.* **19**, 620 (1968).
[20] De John, P. B., Adams, A. D., *Hydrocarbon Process.* **54**, 104 (1975).
[21] Glushchenko, I. M., Melnichuk, A. J., *Khim. Tverd. Topl.* **3**, 123 (1969).
[22] Vans, D. E., *Fuel* **49**, 110 (1970).
[23] Siemieniewska, T., *Koks, Smoła, Gaz* **13**, 33 (1968).
[24] Werner, W., *ibid.* **15**, 36 (1970).
[25] Andrzejak, A., Krajewski, A., *ibid.* **15**, 309 (1970).
[26] Ammosova, J. M., Perederii, M. A., Gorokhova, G. N., *et al., Khim. Tverd. Topl.* **2**, 63 (1973).
[27] Perederii, M. A., *Termookislitel'naya aktivatsya burykh uglei Kansko-Achinskogo basseina* (Thermal oxidative activation of brown coal from the Kansk-Achinsk basin), Thesis, MKhTI Mendeleev, Moscow 1973.
[28] Grebennikov, V. S., Koz'min, G. V., Klimov, O. M., *Khim. Tverd Topol.* **1**, 164 (1973).
[29] Spiehal, W., *Brennst.-Chem.* **41**, No 4 (1960).
[30] Kitagawa, H., *Nippon Kagaku Kaishi* **6**, 1140 (1972).
[31] Ishibashi, K., Noda, Y., Mitsui, S., *J. Feul Soc. Japan* **52**, 336 (1973).
[32] Boikova, G. I., Gribamenkova, T. A., Korchunov, V. S., *Poluchenie, struktura i svoistva sorbentov* (The production, structure and properties of sorbents), Izd. Tekhn. Instituta Lensovet, Leningrad 1975.

[33] Chodyński, A., Sonelski, M., *Koks, Smoła, Gaz* **19**, 256 (1974).
[34] McDonald, D. G., Nguyen, T. G., *Pulp and Paper Magazine of Canada* **75**, 97 (1974).
[35] Siedlewski, J., Rychlicki, G., *Przem. Chem.* **10**, 583 (1969).
[36] Falchuk, V. M., Plachenov, T. G., *Zh. Prikl. Khim.* **17**, 376 (1969).
[37] Plachenov, T. G., Musakina, V. P., Sevryugov, L. B., et al., *ibid.* **17**, 2020 (1969).
[38] Fridman, Z. I., Morozova, A. A., *Khim. Volokna* **4**, 11 (1977).
[39] Chodyński, A., *Koks, Smoła, Gaz* **25**, 72 (1980).
[40] Iley, M., Marsh, H., Rodriguez-Reinoso, F., *Carbon* **11**, 633 (1973).
[41] De Lopez-Gonzales, D. J., Martinez-Vilchez, F., Rodriguez-Reinozo, F., *ibid.* **18**, 413 (1980).
[42] Marsh, H., Iley, M., Berger, J., *ibid.* **13**, 103 (1975).
[43] Andrzejak, A., Janiak, J., Jankowska, H., et al., *Przem. Chem.* **61**, 461 (1982).
[44] Janiak, J., D. Sc. Thesis, University of Poznań, Poznań 1980.
[45] Majowski, G., Toda, T., Sanada, Y., *Carbon* **8**, 681 (1971).
[46] Fica, J., Wałęga, E., Skoczkowski, K., *Koks, Smoła, Gaz* **24**, 222 (1979).
[47] Gąsior, G., Hajewski, P., Fica, J., et al., Polish Patent 111613.
[48] Fica, J., Wałęga, E., Haberski, A., *Proceedings of the VIth All Polish Colloqium on Sorption and the Protection of Man and Environment* Military Technical Academy, Warsaw 1981, p. 157.
[49] Hajewski, P., *Koks, Smoła, Gaz* **27**, 84 (1982).
[50] Chodyński, A., Sonelski, M., Liberacki, J. *ibid.* **26**, 35 (1981).
[51] Morgan, I., Fink, C. E., *Ind. Chem. Eng.* **2**, 219 (1946).
[52] Chodyński, A., Sonelski, M., *Koks, Smoła, Gaz* **22**, 256 (1977).
[53] Kostomarova, M. A., Surinova, S. I., *Khim. Tverd. Topl.* **6**, 3 (1976).
[54] Babel, K., Czechowski, Z., Dębowski, Z., et al., Symposium on Technical problems of protecting respiratory tracts, Proceedings, Rynia, *Biul. Wojsk. Inst. Chemii i Radiometrii* **2**, 158 (1979).
[55] Turonek, M., Zin, M., *Proceedings of the VIth All Polish Adsorption Colloqium on Sorption and the Protection of Man and Environment* Military Technical Academy, Warsaw 1981, p. 163.
[56] Krivovyaz, I. M., *Procesy obrazovaniya prochnykh uglerodistykh materialov* (Processes of obtaining resistant carbonaceous materials), FAN, Tashkent 1970.
[57] Smisěk, M., Černỳ, S., *Active Carbon*, Elsevier, Amsterdam, London, New York 1970.
[58] Hassler, J. W., *Active Carbon*, Chemical Publishing, New York 1951.
[59] USA Patent 2 848 637, 1953.
[60] USA Patent 3 533 961, 1970.
[61] Watari, S., *Kogaku Kogyo* **23**, 366 (1972).
[62] Kostomarova, M. A., Surinova, S. I., Adsorbenty, ikh poluchenie, svoistva i primenenie (Adsorbents, their production, properties and application), Proceedings of All Russian Conference on Adsorbents, *Trudy Vsesoyuzn. Soveshch. po Adsorbentam*, Nauka, Leningrad 1978, p. 58.
[63] Wataru, S., *Kokan* **7**, 19 (1973).
[64] Dębowski, Z., *Koks, Smoła, Gaz* **25**, 280 (1980).
[65] Ignasiak, B. S., Clugston, D. M., Montgomery, D. D., *Fuel* **51**, 76 (1972).
[66] Haberski, A., Heilpern, S., *Koks, Smoła, Gaz* **21**, 289 (1976).

[67] Dębowski, Z., Chodyński, A., Nowak, B., et al., *Proceedings of the VIth all Polish Colloqium on Sorption and the Protection of Man and Environment*, Military Technical Academy, Warsaw 1981, p. 125.
[68] Everett, D. H., Redmann, E., Miles, A. J., *Fuel* **42**, 219 (1963).
[69] Smith, N. R., *Quart. Rev. Chem. Soc.* **13**, 287 (1959).
[70] Jolley, L. J., Poll, A., *J. Inst. Fuel*, **26**, 33 (1953).
[71] Long, F. J., Sykes, K. W., *Proc. Roy. Soc.* London **A193**, 377 (1948).
[72] Reif, A. S., *J. Phys. Chem.* **56**, 785 (1952).
[73] Fedoseev, S. D., *Khim. Tverd. Topl.* **1**, 116 (1972).
[74] Dziubalski, R., Korta, A., Lasoń, M., *Koks, Smoła, Gaz* **23**, 319 (1978).
[75] Donnet, J. B., Couderc, P., Papirer, E., *J. Chim. Phys.* **65**, 1399 (1968).
[76] Japanese Patent 1765 **8**, 1964.
[77] Novikov, I. K., Makhorin, K. E., *Khim. Tverd. Topl.* **2**, 128 (1968).
[78] Kijewska, A., Ciećkiewicz, E., *Przem. Chem.* **52**, 49 (1973).
[79] French Patent 2110554, 1972; British Patent 1302456, 1973 and 1288796, 1972; USA Patent 3677727, 1972 and 3668145, 1972; Australian Patent 414940, 1971.
[80] Dębowski, Z., *Proceedings of the VIth All Polish Colloqium on Sorption and the Protection of Man and Environment*, Military Technical Academy, Warsaw, 1981, p. 120.
[81] Dziubalski, R., Korta, A., Lasoń, M., *Koks, Smoła, Gaz* **21**, 229 (1976).
[82] DIN ISO 787, 11 General methods of test for pigments and extenders; part 11; Determination of tamped volume and apparent density after tamping.
[83] ASTM D 2854-70 Standard Test Method for Apparent Density of Activated Carbon.
[84] DIN 4188 Drahtsiebböden für Analysensiebe.
[85] ASTM D 2862 Standard Test Method for Particle Size Distribution of Granular Activated Carbon.
[86] Ergun, S., Chemical Engineering Progress **48**, 89 (1952).
[87] AWWA B 604-74 Standard for Granular Activated Carbon Sect. 4.6 Abrasion Resistance, 4.6.2 Stirring Abrasion Test, 4.6.3 Ro-Tap Abrasion Test.
[88] ICUMSA (1979) Sugar Analysis, Schneider, F., p. 221 Attrition Hardness.
[89] ASTM D 3802 Standard Test Method for Ball-Pen Hardness of Activated Carbon.
[90] ASTM D 1783-70 (Analysis) Phenolic Compounds in Water.
[91] DIN 19603 Aktivkohle zur Wasseraufbereitung Technische Lieferbedingungen.
[92] AWWA B 600-78 Powdered Activated Carbon.
[93] DAB 7 Deutsches Arzneibuch 7th Edition.
[94] European Pharmacopoeia 'Carbo Activatus', 'Phenazonum'.
[95] ISO 562-1981 Determination of the Volatile Matter Content of Hard Coal and Coke.
[96] ASTM D 2867-70 Moisure in Activated Carbon.
[97] ASTM D 2866-70 Total Ash Content of Activated Carbon.
[98] NORIT — Activated Carbon. Main Grades of Norit Activated Carbon and their Applications, Amersfoort (The Netherlands) 1983.
[99] DIN 53194.
[100] AWWA B 604-74.
[101] ANSI/ASTM D 3466 Ignition Temperature of Granular Activated Carbon.

CHAPTER 3
Structure and Chemical Nature of the Surface of Active Carbon

3.1 MOLECULAR, CRYSTALLINE AND POROUS STRUCTURE OF ACTIVE CARBON

Both the porous structure (texture) and the chemical nature of the surface of active carbon are significantly related to its crystalline constitution. The graphite-like microcrystalline structure is the basic structural unit of active carbon, as in the case of carbon black. The ordering of carbon atoms in an elementary microcrystallite indicates considerable similarity to the structure of pure graphite, the crystals of which consist of parallel layers of condensed regular hexagonal rings spaced 0.335 nm apart. Such interlayer spacing is diagnostic of interaction by means of van der Waals forces. The length of the carbon-carbon bond in individual layers is 0.142 nm. Each carbon atom bonds with the three adjoining ones by means of covalent bonds, and the fourth delocalized π-electron may move freely in a system of conjugated double bonds of condensed aromatic rings. The layers are also so spaced with respect to one another that one-half of the carbon atoms belonging to one layer is placed exactly over or under the centres of hexagonal carbon rings of the adjacent layers. The scheme of arrangement of the carbon atoms in a crystal of graphite is presented in Fig. 3.1.

The formation of the crystalline structure of active carbon begins early during the carbonization process of the starting material. Thus sets of condensed aromatic rings of various numbers, which are the nascent centres of graphite-like microcrystallites, are formed. Although their structure resembles that of a crystal of graphite there exist some deviations from that structure. Thus, among other things, the interlayer distances are unequal in crystals of active carbon and range from 0.34 to 0.35 nm. Again, the orientations of the respective layers generally display

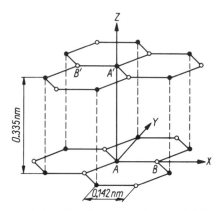

Fig. 3.1 Ordering of carbon atoms in a crystal of graphite.

deviations. Such deviations from the ordering characteristic of graphite, called a turbostratic structure [1], are illustrated in Fig. 3.2. Disordering of the crystal lattice may be caused to a considerable degree both by its defects (e.g. vacant lattice sites) and the presence of built-in heteroatoms. It results from the kind of the raw material used, and the nature and quantity of its impurities as well as the methods and conditions of the production processes of the active carbon. In addition to disordering within the internal structure of the crystallites, the second significant difference between the structure of graphite and that of active carbon lies in the quantity and mutual orientation of the crystallites. The range of order of the crystal structure, which is very high in the case of graphite, is limited for active carbons. As a rule, microcrystallites achieve diameters of between 1 and 10 nm, and usually consist of a number of layers. Their dimensions depend on the type of active carbon, and are greatly

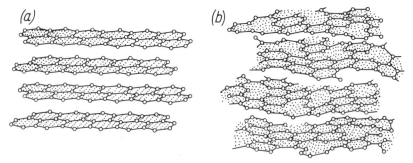

Fig. 3.2 Comparison of three-dimensional crystal lattice of graphite (a) and the turbostratic structure (b) (after Bokros [2]).

influenced by the conditions of production (especially upon temperature and time of heat treatment).

There have been attempts to classify active carbons according to their crystal structure. For example, Franklin [3] proposed the division of carbons into the following two types: those easily undergoing graphitization at high temperatures and those which undergoes graphitization (under the same conditions) only to a small degree. The difference between these two types is well illustrated by comparison of the carbons produced from poly(vinyl chloride) and poly(vinylidene chloride) [3]. Increase of the carbon burn-off temperature from 1000°C to 1720°C for the first type causes the average number of graphite layers in the crystallites to increase from 4.5 to 33 and their diameter from 1.8 to 6.3 nm while for the second type (for a temperature of 2000°C) the number of layers increases from 1.6 to only 2.2 nm. The different abilities to undergo graphitization result mainly from the orientation of the crystallites, as illustrated schematically in Fig. 3.3.

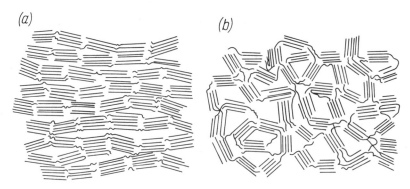

Fig. 3.3 Schematic illustration of the structure of active carbon: (a) easily undergoing graphitization, (b) undergoing graphitization to a small degree (after Franklin [3]).

Carbons featuring a chaotic mutual arrangement of microcrystallites with strong cross-linking between them are characterized by a well-developed porous structure, relatively small true (helium) density (i.e. smaller than 2 g cm^{-3}), considerable hardness, and a small degree of graphitization. In the case of carbon which easily undergo graphitization, the situation is exactly opposite. The orientation of the crystallites is approximately parallel and cross-linkages between them are weak, which facilitates their development by means of addition of whole layers. The porous structure is accordingly poorly developed, the carbons are relatively soft and their density is quite large, approximating to the value typical for graphite and amounting to 2.26 g cm^{-3}. There are also significant differences in other properties, e.g. magnetic susceptibility, number of unpaired electrons, etc.

The above comparison indicates that many of those properties of active carbons, significant as regards their various uses (mainly as adsorbents, catalyst supports, electrode materials) are determined by their crystalline constitution, which depends in turn on the raw material as well as the method and conditions of carbon production. Particularly significant are the degree of development and the nature of their capillary structure. The average active carbons have a strongly developed internal structure (the specific surface often exceeds 1000 and sometimes even 1500 m^2 g^{-1}), and they are usually characterized by a polydisperse capillary structure, featuring pores of different shapes and sizes. Bearing in mind the values of the effective radii and the mechanism of adsorption of gases, Dubinin [4] proposed three main types of pore, namely macropores, mesopores and micropores.

Macropores are those having effective radii $\geqslant$ 100–200 nm and their volume is not entirely filled with adsorbate via the mechanism of capillary condensation (it may occur only for a relative pressure of adsorbate of nearly one). The volumes of macropores are usually in the range 0.2–0.8 cm^3 g^{-1} and the maxima of volume distribution curves according to the radii are usually in the range 500–2000 nm. The values of their specific surface area not exceeding 0.5 m^2 g^{-1} are negligibly small when compared with the surface of the remaining types of pore. Consequently macropores are not of great importance in the process of adsorption as they merely act as transport arteries rendering the internal parts of the carbon grains accessible to the particles of adsorbate. For testing of their properties, mercuric porosimetry is mainly used.

Mesopores, also known as transitional pores, have effective radii falling in the range of 1.5–1.6 nm to 100–200 nm. The process of filling their volume with adsorbate takes place via the mechanism of capillary condensation. For average active carbons, the volumes of mesopores lie between the limits 0.1–0.5 cm^3 g^{-1} and their specific surface areas in the range of 20–100 m^2 g^{-1}. The maximum of the distribution curve of their volume versus their radii is mostly in the range of 4–20 nm. Mesopores, besides their significant contribution to adsorption, also perform as the main transport arteries for the adsorbate. The methods mostly used to test mesopores include adsorption and desorption of gases and organic vapours as well as mercuric porosimetry and testing by electron microscopy.

Micropores have sizes comparable with those of adsorbed molecules. Their effective radii are usually smaller than 1.5–1.6 nm and for average active carbons their volumes usually lie between 0.2–0.6 cm^3 g^{-1}. The energy of adsorption in micropores is substantially greater than that for adsorption in mesopores or at the non-porous surface, which causes a particularly large increase of adsorption capacity for small equilibrium pressures of adsorbate. In micropores, adsorption proceeds via the mechanism of volume filling. For some active carbons, the microporous structure may have a complex nature, e.g. two overlapping microporous structures [5]: firstly one for effective pore radii smaller than 0.6–0.7 nm and termed specific micropores, and the secondly one exhibiting pore radii from 0.6–0.7 to 1.5–1.6 nm termed supermicropores. To characterize the microporous structure of carbons, the adsorption of vapours and gases is primarily applied and, to a lesser extent, the small-angle scattering of X-rays. An example of the full characteristics of distribution of volume for all types of pore versus their effective radii, determined on the basis of adsorption of benzene vapour and mercuric porosimetry, is presented in Fig. 3.4 [6].

Apart from carbons with typically polydisperse porous structure, carbons intended for particular applications and characterized by the prevalence of only one type of pore can also be obtained. They may comprise, for example, almost exclusively microporous carbons (e.g. the increasingly used carbon molecular sieves [7] which enable selective

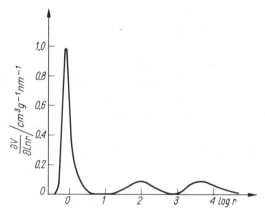

Fig. 3.4 Differential volume distribution curve for pores of active carbon HS-43 versus their effective radii (after Kadlec *et al.* [6]).

separation of specific components of mixtures according to their molecular size) or mainly transitionally porous carbons (the volume of the mesopores may then reach $0.7 \text{ cm}^3 \text{ g}^{-1}$ and their specific surface area $200–450 \text{ m}^2 \text{ g}^{-1}$).

A means of mutual access of these three types of pore is one of the factors of particular relevance to the kinetics of adsorption. There exist a number of different, even totally divergent hypotheses on this subject. For example, according to one of them (developed on the basis of a tree structure) micropores represent branching from mesopores, while another hypothesis views the three types of pores as having direct connection to the surface of the carbon grains. However, these problems remain under discussion.

3.2 NON-CARBONACEOUS ADDITIVES PRESENT IN ACTIVE CARBON

Active carbons usually contain apart from carbon itself, which constitutes generally over 90 per cent of their mass, various types of non-carbon additives differing from each other both in composition and their mode of combination, particularly with the surface. Two basic types of additive can be distinguished: mineral substances occluded in the pores of active carbon (after its combustion) and denoted as ash, and heteroatoms chemically combined with carbon atoms mainly at the edges and corners of elementary planes of graphite crystallites. The type and quantity of each of the two types of additive result from the type of raw material as well as the method and conditions of production of the active carbon.

Mineral additives (ash). The first type of non-carbon additive, i.e. ash, is not chemically combined with the carbon surface. The ash content in various types of active carbon varies over a wide range, depending primarily on the type of raw material, e.g. in the case of carbon produced from plastics or saccharose it does not exceed 0.5–1 per cent, for carbons made from wood the ash content is 3–8 per cent, and if fossil coal is used as a raw material then it amounts to several and even over 20 per cent. The relative ash content also increases with increase in the degree of burning of the coal during activation (moreover, it is usually greater at the exterior of the coal grain). Ash consists mainly of oxides and, in smaller amounts, of sulphates, carbonates, and other compounds of iron, aluminium, calcium, sodium, potassium, magnesium and many other metals. Depending upon the type of raw material, it may comprise different and often fairly large quantities of silicon. The commonly used method of removing ash is to leach active carbon with acids. Due to the complex composition of ash, mixtures of acids, e.g. hydrochloric or hydrofluoric acids, are often used if ash contains substantial quantities of silicon [8].

For most carbons, virtually all the mineral additives are accessible to leaching solutions, which indicates that they occupy some of the open pores of greater radii. In the case of adsorption from the gaseous phase of non-polar adsorbates, e.g. vapours of benzene [9, 10] or argon [11], ash acts mainly as an inert ballast (its porous structure is developed only to a minimal degree). In the examination of processes accompanying the contact of carbons with aqueous solutions, ash can exert some influence on their course to extents depending upon its content of oxides and salts of alkali metals and therefore it is advisable to remove it beforehand.

Heteroatoms chemically combined with the carbon skeleton. Non-carbon additives of the second type are essentially heteroatoms, most usually oxygen, hydrogen, sulphur, and halogens of which the most important is oxygen. This results from its ubiquity and relative ease of combination with the carbon surface, notably during its contact with the mainly non-polar components of solutions. The second type of heteroatom predominating in carbons, i.e. hydrogen as a residue of incomplete carbonization, can be distributed within the whole volume of the active carbon. Incidentally, heteroatoms need not be bonded directly to the carbon skeleton, thus hydrogen may be bonded via a C–OH residue. The number of heteroatoms (especially oxygen) combined with carbon can be changed to a substantial degree, thus a decrease in their content, or even total elimination can be achieved by burn-off of coal at temperatures over 1000°C, and an increase by modification of the surface, e.g. via oxidation. Noting the particularly great influence of the presence of heteroatoms (and oxygen in particular [12]) in active carbon, on its various properties, the problems associated with the chemical nature of the surface will be discussed below in more detail.

3.3 CHEMICAL NATURE OF THE SURFACE OF ACTIVE CARBON

3.3.1 General Remarks

The chemical nature of active carbons significantly influences their adsorptive, electrochemical, catalytic, acid-base, redox, hydrophilic-hydrophobic, and other properties. The great significance of this problem, both as regards its purely cognitive and practical aspects, has made it the subject of much research. However, many problems associated with the chemical nature of the surface are still not entirely explained. Both the growing variety of uses of active carbon and the development of new testing techniques provide the impetus for the continuous increase of interest in testing, the purpose of which is to elucidate comprehensively the chemical nature of the carbon surface. Substantial and detailed considerations of the state of the art in this field can be found in many monographs [12–16] and original papers [17–20]. Aiming to provide a broad view of the subject-matter associated with the chemical nature of the surface of active carbon, we shall discuss here only the seminal problems. It should be emphasized that due to the substantial similarity of the chemical nature of the surface of active carbon and that of carbon black, many aspects equally concern these two carbonaceous materials. As mentioned before, the chemical nature of the surface of active carbon (as well as of carbon black) is determined decisively by the type, quantity and bonding of various heteroatoms, especially oxygen, within it. Heteroatoms may be combined both with peripheral carbon atoms at the corners and edges of crystallites, and in intercrystalline spaces and even in defect zones of particular planes constituting the crystallites. Most heteroatoms are grouped at the surface of active carbon. Apart from their different locations, the heteroatoms are strongly differentiated in terms of their chemical reactivity.

Surface-bound heteroatoms are believed to adopt the character of the functional groups typical for aromatic compounds, and to react in a similar way with many reagents. Heteroatoms bound in the core of the crystalline structure of active carbon are however often virtually inert both because of their very small accessibility and their mode of combination with local atoms. These considerations apply primarily to oxygen, but entirely analogous arguments apply to other heteroatoms, e.g. chemically combined sulphur [13, 18]. It should be added that, as mentioned earlier, surface functional groups often consist of more than one type of heteroatom, e.g. oxygen and hydrogen together as –OH or –COOH.

3.3.2 Surface Functional Groups: Nature and Types

The occurrence of many different types of surface compound has been proved by many independent methods of investigation. Due to the universal occurrence and particular importance of functional groups containing oxygen, they will now be discussed in more detail. The origin of functional groups and processes leading to their formation constitute one of the most significant problems in active carbon chemistry.

Surface functional groups can originate from the starting material from which a particular active carbon is produced. This is especially so for active carbons produced from raw materials relatively rich in oxygen, e.g. wood, saccharose, phenol-formaldehyde resins, following their incomplete carbonization. Substantial quantities of oxygen can be introduced during the production process itself, e.g. during activation of coal by oxidizing gases, such as water vapour and air. Active carbon used predominantly for practical purposes generally includes some percentage by weight of chemically-bound oxygen and a usually much smaller quantity of hydrogen combined with surface carbon atoms either directly or through oxygen. Greater quantities of oxygen can be introduced into the surface of active carbon (just as in the case of carbon black) by deliberately subjecting them to chemical modification. The oxidizing agents used for this purpose can be divided into two groups. The first one consists of gaseous oxidizing agents, such as oxygen, ozone, air, water vapour, carbon dioxide and nitrogen oxides. The second group comprises solutions, particularly of oxidizing materials, the most often used of which are nitric acid, a mixture of nitric and sulphuric acids, hydrogen peroxide, acidic potassium permanganate, chlorine water, sodium hypochlorite and ammonium persulphate. Depending on the type of oxidizing agent used and on the conditions of the modification process, the number of oxygen-containing surface functional groups can differ over a fairly wide range, as can their reactivity. The influence of temperature in the case of oxidizing active carbon with oxygen, may be mentioned here as an example [21]. It should be added that oxygen-containing surface functional groups represent the predominant form of oxygen combined with carbon: they usually constitute about 90 per cent of the total amount of bound oxygen [14]. The occurrence of several types of group, independent of the type of active carbon or how it was modified, is the second essential feature of surface functional groups.

Oxygen surface compounds are usually divided into two main types: functional groups of acidic nature (undergoing neutralization by bases) and basic groups which may be neutralized by acids. The first, or acidic, group is exemplified schematically in Fig. 3.5.

The second, or basic, group is much less well-characterized compared with the first. Usually structures corresponding to chromene [17, 22] or pyrone-like structures [23, 24], illustrated in Fig. 3.6, are attributed to them. It should be noted that the basic properties of

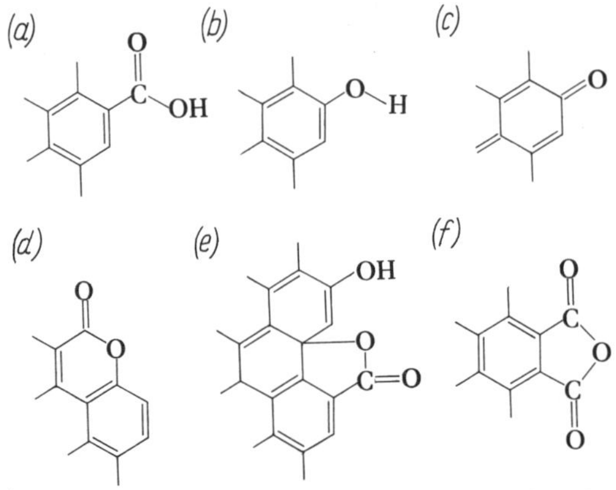

Fig. 3.5 Principal types of acidic oxygen surface functional groups: (a) carboxyl, (b) phenolic, (c) quinonoid, (d) normal lactone, (e) fluorescein-type lactone, (f) anhydride originating from neighbouring carboxyl groups (after Mattson and Mark [15]).

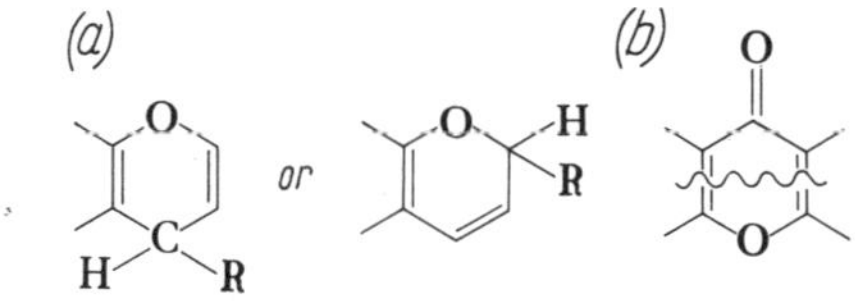

Fig. 3.6 Functional groups of basic character: (a) chromene (after Garten and Weiss [22]), (b) pyrone-like (after Boehm [23]).

particular sites of the active carbon surface are not necessarily associated with the presence of oxygen [25]. Particular types of functional group, both of acidic and basic nature, are described in detail in many publications [12–24]. In order to illustrate the general chemical character of the active carbon surface, the model of a fragment of oxidized active carbon surface, proposed by Tarkovaskya et al. [26], is

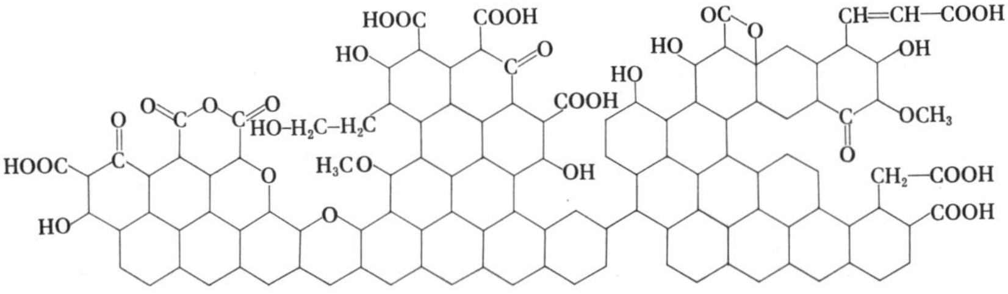

Fig. 3.7 Model of a fragment of an oxidized active carbon surface (after Tarkovskya et al. [26]).

shown in Fig. 3.7. It should be emphasized that oxygen-containing functional groups, even of a given type, can display some, or even fairly substantial, differences in reactivity owing to direct interactions with adjacent groups of the same or other type, or to electronic interactions, via surface carbon crystallites, with more distant groups. The quantity of oxygen combined with the surface in the form of functional groups can not only be increased via oxidation, but also decreased as a result of burn-off of active carbon at temperatures exceeding 1000°C *in vacuo* or in an oxygen-free atmosphere. Under such condition, the oxygen-containing functional groups thermally decompose to yield CO_2, CO, H_2O or H_2 which then undergo desorption.

The chemical nature of the surface of average unmodified active carbon (and also carbon black) is usually intermediate between the two extreme cases discussed here. Only active carbon produced from particular raw materials and by use of specified methods and under controlled conditions of production may provide an exception. Since the chemical character of the active carbon surface is mainly determined by the oxygen combined with it, the identification and quantitative determination of the various types of surface oxygen-containing functional groups, together with their interactions, presents a fundamental problem. Experiments carried out over the last sixty years to elucidate the chemical nature of the surface of active carbon (as well as of carbon black and to a smaller extent graphite, pyrocarbon and other carbonaceous materials) have deployed different methods of investigation, adapted primarily from the field of organic analysis. These approaches are discussed in the next Section.

3.3.3 Methods of Analysis of Surface Functional Groups Containing Oxygen

Since there is a voluminous literature devoted to the characterization of the chemical nature of the surface of active carbons, only the most important methods will be discussed here. These include traditional analytical methods specially adapted for this purpose as well as the newest instrumental methods.

The first group includes all methods which analyse for the effect of reactions proceeding between the surface functional groups and an appropriately selected second reagent present in a contacting liquid solution.

One such method involves selective neutralization of surface acidic groups with bases of various strengths. This method, due to Boehm [27], assumes that the following groups become neutralized as follows: carboxyl groups (with low pK_a) in a solution of sodium bicarbonate, carboxyl and lactone groups in a solution of sodium carbonate, carboxyl and lactone and phenolic groups in a solution of sodium hydroxide, and all previously mentioned groups and the most weakly acidic groups, i.e. carbonyl, in a solution of sodium ethoxide. It should be emphasized that this approach is to some extent simplified since it does not take into account long-range interactions between functional groups through conjugated π-systems which can influence acidity to some extent. Alternatively, potentiometric alkalimetric titration [28–30], using one strong base, e.g. sodium hydroxide can also be applied. From analysis of the shape of the titration curves, the contribution of functional groups of various acidities can be determined. Similarly, other bases can be utilised, e.g. barium hydroxide [18], to determine carboxyl groups in close proximity; acetates, e.g. of sodium or calcium, are also used.

Conversely, for quantitative estimation of the basic centres of a carbon surface, neutralization with dilute hydrochloric acid is investigated [31].

In experiments aimed at more unequivocal identification, as well as selective quantitative estimation, of particular types of oxygenated functional groups, reactions developed from organic chemistry are used. Such methods involve transformation of the target groups by specific reagents (usually used in excess) and then determination of the concentration of transformed groups or other products formed in the course of reaction. Such transformations include acetylation, methylation, reactions with hydroxylamine, diazomethane, methylmagnesium iodide and lithium aluminium hydride. An extensive survey of the reactions of oxygen-containing functional groups of acidic nature is provided in Donnet's review [19].

In the case of surface functional groups displaying redox behaviour, e.g. quinones, redox reactions with the ions of some metals can be utilised, e.g. Sn^{2+}, Ti^{3+}; conversely, hydroquinone groups can be detected by reducing Fe^{III} to Fe^{II}. The oxidizing properties of a carbon surface can be determined [32–34] by contact with a solution of potassium iodide in N,N-dimethylformamide (DMF) according to the equation:

$$3I^- = I_3^- + 2e^-. \qquad (3.1)$$

The originators of this method believe that electrons formed in reaction (3.1) are used for reduction of surface oxides according to equation (3.2):

$$C_xO + e^- = C_xO^- \tag{3.2}$$

with the resulting anionic form of these oxides associated with an equivalent quantity of potassium cations. It can be assumed that practically all the I_3^- ions originating from reaction (3.1) remain in solution and can be determined by titration with aqueous sodium thiosulphate. This follows from the large value of K for equation (3.3):

$$I_2 + I^- = I_3^- \tag{3.3}$$

amounting to log $K = 7$ in DMF solutions, thus the iodine produced appears practically exclusively as I_3^-, the adsorption of which at a carbon surface is negligible. Thus the quantity of these anions produced from 1 g of active carbon (or carbon black) can be considered, according to Vignaud and Brenet [32–34], as a measure of the 'oxidizing capacity' of the carbon surface.

In the course of more detailed research into this process [35, 36], the interpretation given previously has been partly questioned, and a model describing the observed phenomena quantitatively has been developed. It has been assumed that the system reaches a state of equilibrium when the potential of carbon versus solution reaches the value equal to the redox potential corresponding to reaction (3.1). The value of this potential can be calculated on the basis of the simplified Nernst equation in the following form:

$$E = \frac{RT}{2F} \ln \frac{z}{(m-3z)^3} \tag{3.4}$$

where R is the universal gas constant, F is Faraday's constant, T is the temperature, z equals the molality of I_3^- anions formed in solution, and m is the initial molality of the solution of potassium iodide in DMF. In equation (3.4) we neglect activity coefficients and adopt a scale of potentials in which the value of the standard potential for the couple I_3^-/I^- in DMF is zero. On the other hand, the charge of the carbon potential from a certain initial value E_0 (corresponding probably to a zero point of the charge on carbon) to the final value specified by equation (3.4), is connected with the removal of a particular charge into carbon via an equation of the following form:

$$E = E_0 - \frac{2 \times 10^{-3}}{\bar{C}} ZF \tag{3.5}$$

where: Z is the oxidizing capacity in mmol I_3^- per gram of carbon ($Z = z\, m'_s/m'_c$, m'_s is the mass of solvent, and m'_c — the mass of carbon), $\bar{C}$ — the average capacity of the electrical double layer with reference to 1 g of carbon. Equation (3.5) presents the linear dependence of the oxidizing capacity Z on potential E which can be written in the form:

$$Z = -aE + b \tag{3.6}$$

where the angular coefficient $a = \bar{C} \times 10^3/2F$ is proportional to the specific surface area of carbon, and $b = aE_0$ (or E) should characterize the state of the carbon surface. The procedure consists here in determining several (a dozen or so) experimental points of chosen m, m'_s and m'_c values, estimating the value of z, plotting the relationship $Z = f(E)$ and determining on this basis the values of a and b. That this model is realistic is indicated by the fact that the capacities of the electrical double layer determined from it and amounting to 14 μF cm^{-2} approximates to the values obtained for different carbonaceous materials by other authors who used other, independent methods (for instance, cyclic voltammetric measurements [37, 38] or loading curves [39–41]. It has also been shown experimentally [42] that for N-220 type carbon black modified by various methods, a linear relationship occurs between E and the content of oxygen-containing surface functional groups undergoing neutralization by a sodium hydroxide solution. On the basis of a wealth of research material it has also been ascertained [43] that the substitution of DMF by other aprotic solvents with similar properties such as dimethyl sulphoxide (DMSO) or acetonitrile (AN) leads to conclusions analogous to those drawn in the case of DMF (Fig. 3.8); the equation $Z = f(E)$ always gives a linear plot, while its slope depends on the kind of solvent used. These differences are due to the varying capacities of the double electrical layer at the carbon-electrolyte solution boundary for different solvents (the ratio of coefficients a_{solv}/a_{DMF} being constant irrespective of the carbon involved). Also the difference between the respective values E_0 determined for two different solvents is constant and independent of the kind of carbonaceous material (active carbon, carbon black) used. The studies have shown that DMF is the solvent best meeting the requirements of the method of testing carbons (best accuracy of analytical determinations).

Recapitulating the methods already discussed, we should note that the use of several different methods, based on different reactions (different reagents), for determining a particular type of functional group often leads to divergent results. It is impossible therefore to obtain an

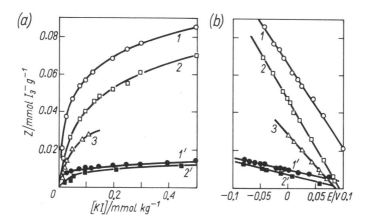

Fig. 3.8 Dependence of the oxidizing capacity Z of active carbon AG-3 (1–3) and, for comparison, of Vulcan 6 carbon black (1'–2') on: (a) the initial concentration of KI in the solution, and (b) the equilibrium redox potential of the solution. The solvents used are: DMF (1, 1'), DMSO (2, 2') and AN (3) (after Mioduska et al. [42]).

absolutely precise description of the chemical make-up of the surface of carbons solely on the basis of these methods. Usually independent methods, based on quite different principles, are also applied, e.g., advantage is taken of the quite significant differences in the thermal stability of particular kinds of surface functional group. These differences are due to the varying energies necessary to split particular chemical bonds, e.g. these are much smaller in the case of C-carbon skeleton bonds for carboxylic groups as compared with quinone or phenolic groups. Thus, in the course of heating a carbon (or carbon black) sample under vacuum or in a neutral gas, particular surface compounds decompose at different characteristic temperatures [44–47] yielding such products as the following:

(a) Carbon dioxide — from decomposition of carboxylic and lactone groups — in the range from *ca.* 200 to 700–800°C.

(b) Carbon monoxide — from decomposition of quinone, phenol and ether groups — in the range from *ca.* 500 to *ca.* 1000°C.

(c) Water — from decomposition of phenolic groups — in the range from 200–300 to 400–500°C, and

(d) Molecular hydrogen — from recombination of hydrogen atoms liberated as a result of splitting of C–H and O–H bonds — in the range above 500–700°C (up to 25% of the initial hydrogen may remain bonded to carbon even at 1200°C).

Investigation of the above-mentioned processes providing information on the chemical nature of the active carbon surface, as well as of carbon black, is conducted by several different methods. One such method involves qualitative and quantitative analysis of the decomposition products of the surface compounds evolved on heating the carbonaceous sample at one, several or a dozen or so selected temperatures (usually up to 1200°C) [44–46]. More and more frequently a thermal desorption method is used [48] with a programmed increase of temperature accompanied by simultaneous analysis of the evolved decomposition products by instrumental techniques such as mass spectrometry or gas chromatography. In this method known as pyrolytic volatile analysis (PVA), changes of the rate of evolution of the products, evident as maxima on the thermograms at characteristic temperatures, are recorded during the programmed heating of the carbon samples. The temperature dependence of the evolution rate is affected both by the type and amount of the particular surface compounds, and their immediate neighbourhood. Apart from methods in which the gases evolved during the pyrolysis of carbonaceous materials are analysed, are the increasingly common studies of the changes that the samples undergo on heating. We are referring here to the derivatographic investigations that allow us to determine the variation (decrease) of mass of these samples as a function of temperature when increased in a programmed way with simultaneous determination of the thermal effects accompanying the decomposition of various kinds of functional groups [49, 50]. Very good results are achieved by combining this method with the analysis of the evolved gases [51].

In general it appears that extensive studies of the thermal decomposition of the surface compounds yield valuable information on their structure, which provides an important supplementation of results obtained by other, independent methods. The interpretation of the relations obtained is almost always difficult. This is due primarily to the different behaviour of the given functional groups depending on whether they are isolated or, in the case of a greater degree of surface coverage with oxygen, they interact directly with similar or other groups in their immediate neighbourhood (yielding, e.g., acid anhydrides or lactones). In effect this gives a very complicated, multistep thermal decomposition, and the limits of the temperature ranges characteristic for particular processes are usually rather diffuse.

A common feature of the methods considered hitherto is the direct participation of the oxygen-containing surface functional groups in chemical reactions when they undergo, under different conditions, different conversions (ion exchange, derivatization, oxidation-reduction, etc.) or destruction (thermal decomposition). The other, common research methods do not lead to changes in the chemical structure of the oxygen-containing surface functional groups. By way of example we can mention here the method of determining the heat of wetting the carbon by water which provides information on the hydrophilicity of the carbon and the degree of oxidation of its surface. It has been shown [52, 53] for many samples of active carbons subjected to various kinds of thermal treatment that a linear relationship exists between the heat of wetting and the amount of oxygen-containing surface compounds that release carbon dioxide during pyrolysis. Such a relationship, however, does not extend to the total quantity of oxygen bonded to the carbon surface. The heat of wetting with water can also be correlated with the quantity of base undergoing neutralization on contact of basic solutions with the carbons [52–54]. Again, the possibility of obtaining important information from the thermal effects accompanying contact of the active carbon surface with aqueous solutions of potassium hydroxide should be noted [55].

The use of these relationships as research method may be very useful as supplementary to other, independent methods.

Methods based on instrumental techniques used in chemical analysis find direct application to the chemical structure of the surface of active carbons. These fall into a separate group which includes, among others, the electrochemical methods among which the most important are polarography, voltammetry and currentless potentiometry.

The first of these methods is the *polarography* (usually with a dropping mercury electrode) of suspensions of ground active carbon in aqueous or non-aqueous solutions of the base electrolyte. The aim of these studies is to assign the determined half-wave potentials to the oxidation-reduction reactions of particular surface functional groups by analogy with the behaviour of known organic substances in the same systems. To confirm the conclusions based on such comparisons, the functional groups subject to a particular study are derivatized by some selective reagent. In some investigations a series of samples reflecting varying amounts of oxygen bonded to the surface of the same active carbon have been carried out. The chemical similarity of the surfaces of active carbons and carbon black means that conclusions based on studies of one of these are often found to be applicable to the others.

Selected examples of polarographic studies of the chemical structure of carbon surfaces are described below.

In an early work [56], in which a suspension of carbon black in N,N-dimethylformamide with tetrabutylammonium iodide as base electrolyte was studied polarographically, the cathodic wave was assigned to the reduction of quinone groups, while the anodic reaction referred to the oxidation of hydroquinone groups. If, prior to the measurements the carbon black was treated in the first case either with lithium aluminium hydride or methylmagnesium iodide and in the second case either with methylmagnesium iodide or diazomethane, then both polarographic waves disappeared. A similar effect was obtained if carbon black was heated at 1950°C. These facts confirm that quinone and hydroquinone groups occur on the carbon surface. These findings are supported in another study [57] in which suspensions of active carbon (and of other carbonaceous materials) in potassium chloride and sodium hydroxide solutions were analysed. In later works [58–61] polarographic studies of active carbons provided further information on the possibilities of identifying by this method various oxygen-containing surface functional groups. It has been shown for suspensions of modified active carbons (demineralized and oxidized by various methods), that the half-wave potentials obtained in the cathodic range, corresponding to the reduction of particular groups, occur at -0.5 to -0.6 V, -1.35 V and -1.60 to -1.65 V for quinone groups, carbonyl groups, and carboxylic groups, respectively. It has also been shown [58] that the amplitudes of the polarographic waves are proportional to the content of carbonyl groups (on the surface of samples representing different degree of oxidation) as determined quantitatively from their reaction with hydroxylamine hydrochloride.

These various examples of polarographic analysis applied to the surface of active carbons indicate that this method can provide information, primarily qualitative, about the surface functional groups participating in redox reactions. Particularly valuable results may be obtained if, in addition, analytical determinations or certain instrumental methods (e.g. infrared spectroscopy [56]) are applied. The polarographic method is being developed to this end and may well find wider application in the quantitative determinations of particular functional groups.

The second most important electrochemical method (involving current flow) used in studies of the surface of carbonaceous materials is the *cyclic voltammetry* (*potentiodynamic*) *method*. Before we consider this method it should be noted that the identification of oxygen-containing surface groups by electrochemical methods involving current flow presents difficulties related to the interaction of these groups with the substrate (graphite-like crystallites) in which the π-electrons of the condensed multiring systems are displaced. These interactions, as well as those between particular surface groups, may perturb their properties. Other difficulties are caused by the porous structure of the active carbons which produces a non-uniform distribution of the potential, and hence non-uniformity of the electrode processes at different sectors of the surface. The charge fed to the electrode is used for neutralizing the ohmic resistance of the carbon electrode, charging the electrical double layer, and activating the surface Faraday processes (which are the basis for interpreting the processes proceeding on the electrode).

The cyclic voltammetry method was used, for example [62], to study oxygen-containing compounds on the surface of furnace black: this was unmodified, fired at 1000 and 2700°C (to obtain various degrees of graphitization) and then either electrochemically oxidized in 96% phosphoric acid, oxidized chemically with chromic acid or oxidized with aerial oxygen at 500°C. The authors observed the presence of the quinone-hydroquinone redox system for carbon electrodes (made from the furnace carbon samples in 1.0 molar sulphuric acid) at a potential variation rate of 14 mV s^{-1} in a potential range of 0.05–0.18 V. The surface concentrations of the oxygen-containing groups determined by graphical integration of the current-potential curves were in the range 10^{-11}–10^{-10} mol cm^{-2} (oxidation of the furnace black having produced an increase in their content, and graphitization a decrease) which accords with the results of chemical studies of carbon black obtained by other authors. When the cyclic voltammetric method was applied to active carbon [38] in sodium hydroxide and sulphuric acid solutions of concentrations 1 mol dm^{-3} and 0.5 mol dm^{-3}, respectively, at a potential variation rate of 0.3×10^{-3} V s^{-1}, the curves obtained had a shape similar to those for furnace black and carbonaceous materials; both cathodic and anodic parts featured a single maximum in its curves for active carbons anodized in sulphuric acid (0.5 mol dm^{-3}). The potentials corresponding to those maxima had a value of 0.625 V.

One may conclude that, in general, active carbons show a distinct similarity to various other carbonaceous materials such as carbon black or glassy carbon as regards their electrochemical properties (the shapes of the cyclic voltammetric curves are determined by similar oxidation and reduction processes).

Recently [63] constant and alternating current voltamperometry at a hanging mercury drop electrode (HMDE) was applied to active carbons dispersed in electrolyte solutions. These focused on the phenomena taking place during adhesion and accumulation of carbon particles on the mercury surface and during reduction of electroactive oxygen present on the carbon surface. This method clearly has potential for estimating the effects of chemical modification of active carbon.

The possibility of using the gold amalgam rotary disc electrode for studying carbon suspensions in electrolyte solutions should also be noted [64].

The testing of carbonaceous materials by the electrochemical methods discussed here continues to develop in view of the ever wider application of these products as electrode materials in various chemical sources of electric current [65].

The last electrochemical method considered here is based on measurements of the potential of the carbon-electrolyte solution boundary under current-free conditions. The suggested procedure [66], which is a modification of that developed earlier by Tomassi *et al.* [67–69], consists of determining the time-dependence of the potential of ground active carbon immersed in aqueous potassium chloride solution (0.5 mol dm^{-3}) (with respect to the saturated calomel electrode). The carbon suspension in the solution is stirred during measurements by a stream of gas (nitrogen, air or oxygen), and electrical contact is ensured by an immersed platinum electrode with which the carbon particles collide. Theoretical aspects of these measurements are considered in detail by Tomassi and Ufnalski [70]. The time-dependence of the carbon-solution potential makes it possible to estimate fairly rapidly reliable values of the potentials (which is usually possible after several hours of contact of the carbon with the solution). Several simultaneous processes contribute to the establishment of the value of the potential, which is usually a complex quantity or 'mixed potential'. One of these is the electroreduction of oxygen supplied from the solution by diffusion and partially adsorbed on the carbon surface which leads to the generation of hydroxide ions in the double layer, and further, owing to ion exchange, to alkalization of the solution. This process is considered in more detail later in the text. Another process, related to the chemical structure of the carbon surface, is of practical importance for testing this surface. It has been found that the acid-base properties of the carbon surface play a dominant role in the process [71]. Acidic functional groups participate in ion exchange reactions with the electrolyte cations according to the stoichiometric equations:

$$C_x-COOH+K^+ = C_x-COOK+H^+ \tag{3.7}$$

or

$$C_x-OH+K^+ = C_x-OK+H^+ \tag{3.8}$$

where C_x is a carbon atom at the surface. Eventually this leads to a lowering of the solution pH which in turn causes a significant increase of the potential (oxygen electrode process in acidic solution). It has been found that the more oxidized is the carbon, the smaller the differences of the potentials as determined under air and nitrogen atmospheres. This is due both to the presence of a large number of acidic functional groups and to oxygen adsorbed in various forms. It has been shown that there is a quantitative relationship between the difference of potentials and the ratio of the number of acidic to basic groups on the surface. The potentiometric method, thanks to its great simplicity and the relatively short time of the measurements, may be very useful, e.g., for evaluating the effects of chemical modification of the active carbon or carbon black surface [72].

Another group of methods for examining the chemical nature of the surface of active carbons (and also carbon blacks) is based on IR spectroscopy [53, 73–80]. Though the low accuracy of the results obtained does not enable, in principle, the use of these methods in quantitative determinations, they usually provide many significant qualitative data regarding the kind of functional groups present on the carbon surface. Transmission IR spectroscopy has been utilised, in which the carbon samples are made up in the form of pastes with nujol [56, 73], or pellets pressed with potassium bromide [74, 75], and also reflection IR spectroscopy [76–78]; more recently transmission IR spectroscopy of carbonaceous materials in the form of films [79, 80] and IR photothermal beam diffraction spectroscopy [81, 82] have been used. The spectra have provided clear evidence of the presence on the surface of the particular kinds of oxygen-containing functional groups that have been identified independently by other methods. The use of IR spectroscopy (especially in the case of carbon films [74]) has also enabled investigation of the processes of carbonization, activation and chemical modification of the carbon surface. It has also facilitated explanation of many important aspects of the adsorption mechanism (e.g. the detailed interaction of some adsorbate molecules with the carbon surface and the functional groups bonded to it).

An instrumental method recently used quite often is electron paramagnetic resonance (EPR) [30, 83–85]. The EPR method is used both for determining the number of paramagnetic centres on the active carbon surface as a function of their origin and method of production and modification, and for studying the interaction of the carbons with oxygen (via its various types of bonding with the carbon surface).

One should note that the EPR results of different authors are often difficult to compare in view of the great diversity of forms in which oxygen is combined with the carbon surface, the importance of the preliminary treatment, and the role of contaminants present in the carbon. Thus the EPR method does not provide information on the chemical structure of the carbon surface that enables practical generalization. However, this method can be valuable in certain well-defined examples.

3.4 EFFECT OF THE CHEMICAL NATURE OF THE ACTIVE CARBON SURFACE ON ITS ADSORPTION PROPERTIES

3.4.1 General Remarks

The chemical nature of the surface of active carbon is regarded as the most important factor, apart from the porous structure, that determines the adsorption properties. The development of the porous structure (pore volume, specific surface area) plays the major role in adsorption from the gas phase. The chemical nature of the adsorbent surface is in this case of minor importance (its role is most apparent for polar adsorbates). In adsorption from the liquid phase, however, the situation is quite different. We should consider here separately aqueous and non-aqueous, and electrolyte and non-electrolyte solutions. In adsorption from the liquid phase, the role of the chemical nature of the surface increases significantly relative to the porous structure, and in some cases dominates. In view of this we shall discuss separately those cases in which the effect of the chemical nature of the surface is specially significant.

3.4.2 Adsorption of Polar Adsorbates from the Gas Phase

Investigations of the adsorption of polar adsorbates from the gas phase onto active carbons involve both the adsorption process itself and its susceptibility to the properties of the active carbons. Usually analysis of adsorption and desorption isotherms or heats of adsorption enables conclusions to be drawn regarding interactions between the carbon surface and the adsorbate. Among the adsorbates used most frequently for this purpose are water vapour, vapours of alcohols and amines and gaseous ammonia and sulphur dioxide [30]. Discussion will be focused on the adsorption of water vapour (the adsorbate used most commonly). In this type of investigation the adsorption and desorption isotherms are usually determined for a series of samples of one or several kinds of carbon which have been subjected to chemical modification. By comparing the adsorption isotherms many significant conclusions may be drawn as regards changes in the chemical nature of the carbon surface consequent upon the method and conditions of the modification

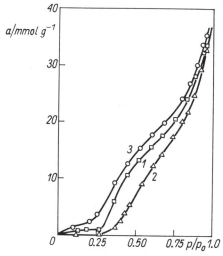

Fig. 3.9 Adsorption isotherms of water vapour (20°C) on active carbon CWN-2 samples: 1 — unmodified, 2 — heat treated under vacuum at 500°C, 3 — oxidized with aerial oxygen at 400°C (after Świątkowski [87]).

process. Particularly useful for this purpose are the initial sections of the isotherms, which refer to relatively low equilibrium pressures of the adsorbate. For a selected pressure, the adsorption found increases with the number of oxygen centres bonded to the surface of the particular sample [86, 87], as illustrated in Fig. 3.9.

The following equation has been proposed [88] to describe the initial section of the adsorption isotherm of water vapour:

$$a = a_0 ch/(1-ch) \tag{3.9}$$

where: a is the adsorption, a_0 is the content of primary adsorption centres, c equals the ratio of the kinetic constants, and h is the relative equilibrium pressure.

Equation (3.9), known as the Dubinin–Serpinskii (DS-1) equation, satisfactorily describes the adsorption of water vapour only when $h < 1/c$. At higher pressures deviations occur since, as the adsorbed water fills the pores, the number of active centres participating in adsorption decreases. Dubinin and Serpinskii took this into account [89] by assuming that this decrease is proportional to $1 - ka$ (where k is a constant) and advanced an equation known as DS-2 which gives a better description of the adsorption of water vapour at higher relative pressures:

$$h = \frac{a}{c(a_0 + a)(1 - ka)} \qquad (3.10)$$

The problems associated with modelling the adsorption of water vapour are discussed in greater detail in review [90]: that the situation remains fluid is due to the constant appearance of new papers on this subject, e.g. [91].

Further detailed studies of sorption on modified carbons and carbon blacks have shown that adsorption of water vapour is not proportional to the total quantity of oxygen bonded to the carbon surface, but only to the content of those oxygen-containing groups which release carbon dioxide on thermal decomposition. It has been found that for each functional group of this kind there is approximately one adsorbed water molecule which is bonded to that group, most probably by a hydrogen bond [52, 92]. This bond is so strong that irreversible adsorption may occur in this case (the hysteresis loop remains open even for very low pressures) [52]. In view of the heterogeneity of the surface and different kinds of interaction between the functional groups it is impossible to determine quantitatively, with sufficient accuracy, the content of acidic functional groups from analysing the shape of the isotherm; it is possible, however, to evaluate successfully from water vapour adsorption the effects of chemical modification of the carbon. It should also be noted that adsorption of water vapour may also be successfully utilized for investigation of modified carbon samples whose surface also contains heteroatoms other than oxygen, e.g. sulphur [93].

3.4.3 Adsorption of Non-Electrolytes from Binary Solutions

The shape of the adsorption isotherms for binary liquid solutions when one of the components is a polar and the other a non-polar substance (e.g. alcohol-benzene or alcohol-carbon tetrachloride systems) depends largely on the chemical nature of the surface. From the results of investigation [52, 94, 95] of a range of active carbon and carbon black samples with modified surfaces (oxidized with various oxidizing agents or heated under vacuum at different temperatures), it has been found that the shape of these adsorption isotherms is affected significantly not only by the number but also by the kind of the oxygen surface compounds. It has been ascertained that the greater the content on the surface of functional groups of strongly acidic character (releasing carbon dioxide on thermal decomposition and neutralized even by weak bases), the greater is the degree of adsorption of the polar adsorbate, e.g. methanol or ethanol. Such oxygen surface compounds provide for polar adsorbates centres with which they may combine via hydrogen bonds. Removal of the acidic functional groups from the surface (due to their thermal decomposition) produces a distinct decrease of adsorption of the alcohols to the advantage of the non-polar adsorbent, e.g. benzene. The sites at which the latter may be adsorbed (thanks to π-electron interactions) are the free crystallite surfaces and the carbonyl groups. To illustrate the influence of the chemical nature of the active carbon surface on adsorption from binary non-electrolyte solutions, isotherms of adsorption from ethanol-benzene solutions on modified active carbon samples with different quantities of oxygen bonded to the surface (0.35, 2.05, and 6.10 wt.%) are shown in Fig. 3.10. It should be noted that from the relative position of the isotherm one can draw conclusions as to the direction and degree of change of a number of functional groups (mainly acidic) due to chemical modification of the active carbon, although of the method and

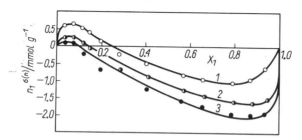

Fig. 3.10 Isotherms of adsorption from binary ethanol-benzene solutions on samples of active carbon CWZ-3: 1 — oxidized with nitric acid, 2 — unmodified, 3 — heated in nitrogen at 1100°C (after Jankowska et al. [95]).

parameters of the modification process do not greatly influence the porous structure and specific surface area of the carbon samples.

3.4.4 Adsorption of Electrolytes, and Electrode Properties of Active Carbons

Adsorption of electrolytes on active carbons from aqueous solutions has been for about sixty years the subject of numerous research works with the aim of establishing of the mechanism of this process. Among many theories developed in this period to describe the interactions of carbon surfaces with electrolyte solutions, the best experimentally- and theoretically-based is the electrochemical theory of adsorption advanced by Frumkin et al. [39, 96–99]. Studies on these problems were started by these authors before 1930 when many other studies were being published in this field. However, these various studies failed to give a unified explanation of all the experimental observations. The same applies to later theories [17, 22, 100] which succeed only as regards certain aspects of the whole problem. Detailed comprehensive discussion on this topic can be found in various monographs, e.g. [15]. A reasonably full interpretation of the phenomena accompanying the contact of active carbon with an electrolyte solution is given by the Frumkin electrochemical theory which is generally accepted by most researchers in this field. We shall therefore give it considerable attention.

According to the Frumkin theory, the process of adsorption of electrolytes on active carbon is strictly related to the potential drop at the carbon-solution boundary and the capacity of the electrical double layer. The magnitude of the potential drop depends in turn on the quantity of electrochemically active gases adsorbed on the carbon surface, as well as on the polarization of the carbon electrode on application of a potential. If for example oxygen adsorbed at a low temperature is present on the carbon surface, then the carbon behaves like an oxygen electrode in which hydroxide ions formed in aqueous solutions make up the inner boundary of the electrical double layer:

$$C_xO + H_2O \rightarrow C_x^{2+} \ldots 2OH^- \qquad (3.11)$$

where C_x are surface carbon atoms.

In this case carbon is charged positively with respect to the solution and may exchange OH^- ions for anions A^- present in solution, *viz.*

$$C_x^{2+} \ldots 2OH^- + 2H^+ + 2A^- \rightarrow C_x^{2+} \ldots 2A^- + 2H_2O \qquad (3.12)$$

or for salts:

$$C_x^{2+}\ldots 2OH^- + 2K^+ + 2A^- \to C_x^{2+}\ldots 2A^- + 2K^+ + 2OH^- \quad (3.13)$$

where A^- is the anion, and K^+ the cation.

Active carbons usually have a large specific surface area due to which the electrical double layer usually possesses a significant capacity which, in turn, leads to a marked anion-exchange capacity. The magnitude of the positive potential may be increased by treating the carbon with ozone or coating its surface with platinum. Adsorption of acid on the surface of active carbon may also proceed [98] in the presence of oxygen via a different mechanism:

$$C_xO_2 + 2H^+ + 2A^- \to C_x^{2+}\ldots 2A^- + H_2O_2 \quad (3.14)$$

The form in which oxygen is bonded to the carbon surface determines whether water or hydrogen peroxide is formed in the reaction.

If hydrogen is adsorbed on the surface of (platinum coated) active carbon, then it behaves like a hydrogen electrode in which the internal boundary of the electrical double layer is formed by hydrogen ions according to the scheme:

$$C_x + H_2 \to C_x^{2-}\ldots 2H^+ \quad (3.15)$$

The negatively charged carbon acquires in effect ion-exchanging properties with respect to the cations present in solution. The bonding of the cations proceeds as follows:

$$C_x^{2-}\ldots 2H^+ + 2K^+ + 2A^- \to C_x^{2-}\ldots 2K^+ + 2H^+ + 2A^- \quad (3.16)$$

and it is accompanied by acidification of the solution.

This mechanism of adsorption of electrolytes on active carbon has been confirmed subsequently many times by other authors [101, 102].

Though the Frumkin theory gives the most comprehensive explanation of the interaction of ions with the carbon surface, it does not provide a full interpretation of the accompanying phenomena. We shall consider here two of the problems arising in this connection.

The first is the specific adsorption of ions. This accompanies ion-exchange adsorption to various degrees depending on the kind of anion present in the electrolyte. The use of potentiometric measurements under current-less conditions (see para 3.3.3) made it possible to explain in the case of active carbon immersed in a halide solution, both the mechanism of specific adsorption and the contribution of this to the total adsorption of anions [103]. It has been ascertained that for equal magnitudes of total adsorption of the ions, Cl^-, Br^-, and I^-, the potentials of the active carbon-electrolyte boundary differ significantly and decrease in the same order. It is also known that the electron affinity of the halogens decreases in the same direction. On this basis it has been suggested that a complex is formed on the carbon surface featuring partial charge transfer, the ion being the donor and carbon the acceptor of electrons, *viz.*

$$C^+ \;|\; \overset{\delta-}{\curvearrowleft} X^-$$
$$C^+ \;|\; \overset{\delta-}{\curvearrowleft} X^-$$
$$C^+ \;|\; \overset{\delta-}{\curvearrowleft} X^-$$

where: $X^- = Cl^-$, Br^- or I^-.

The above suggestion is in full agreement with the results of independent studies [104] on the specific adsorption of halide anions on mercury. The approximate values of the coefficients of partial charge transfer determined for the adsorbed ions are 0.2, 0.4 and 0.7 for Cl^-, Br^- and I^-, respectively. It should be further noted that ion exchange between the OH^- ions present in the double layer and the anions present in solution also depends on the properties of these ions. This exchange process proceeds more readily, the easier the given ion undergoes dehydration. The heat of hydration of the ions $-\Delta H_h^i$ (where h refers to hydration and i is the hydrated ion) may be used as a convenient parameter for comparisons [105].

Attempts have also been made to determine the mechanism of the specific and non-specific adsorption of anions by the SCF LCAO MO method in the modified approximation of Boyd-Whitehead [106]. In the calculations the perinaphthene radical cation was used as the model cation.

The second problem is the effect of the chemical nature of the active carbon surface on anion adsorption. In Table 3.1 the results are summarized of investigations of the adsorption of Br^- and Cl^- anions and Na^+ cation on a range of active carbon samples with a widely varying contents of surface-bonded oxygen (due to oxidation with aerial oxygen or heat treatment *in vacuo* at different temperatures). From analysis of these experimental data it follows that adsorption of anions increases with that of cations as the content of oxygen-containing acidic functional groups increases. As one of the main causes of this we may surmise that increased levels of surface oxidation enhances the number

Table 3.1

Content of oxygen-containing surface functional groups and adsorption of ions on samples of active carbon CWN-2 (after Świątkowski [87])

Conditions of modification of carbon		Content of functional groups /mmol g^{-1}					Molar ratio of total acidic groups to basic group contents	Adsorption from NaBr(NaCl) solution of concentration 2 mol kg^{-1} solvent	
		COOH—	—COO—	—OH	=CO	Basic		$a_{Br^-}(a_{Cl^-})$ mmol g^{-1}	a_{Na^+} mmol g^{-1}
Heating under vacuum for 24 h	500°C		total content		0.063	0.32	0.197	0.0023	0.0014
	300°C							0.0412	
	200°C							0.0985	
Unmodified carbon		0.04	0.03	0.11	0.03	0.205	1.02	0.189 (0.0508)	0.0858 (0.0205)
Oxidation with aerial oxygen for 6 h	200°C	0.09	0.07	0.12	0.03			0.211	
	300°C	0.27	0.17	0.26	0.03			0.272	
	400°C	0.90	0.29	0.78	0.03	0.087	23.0	0.361 (0.119)	0.392 (0.153)

of strongly acidic functional groups; the protons of the latter may undergo exchange with the solution cation producing strong acidification of the solution and consequently a significant increase of the potential favouring the adsorption of anions [107]. On the other hand surface oxidation of the carbon leads to its increased hydrophilicity which facilitates considerably penetration of the porous structure by aqueous electrolyte. It should be noted here that by suitable chemical modification of the carbon surface (not as regards oxidation but also incorporation of nitrogen atoms), carbons may be obtained with designed ion-exchange properties. These problems are discussed in detail in ref. [108].

References

[1] Biscoe, J., Warren, B. E., *J. Appl. Phys.* **13**, 364 (1942).
[2] Bokros, J. C., in: *Chemistry and Physics of Carbon,* P. L. Walker Jr. (Ed.), vol. 5, M. Dekker, New York 1969.
[3] Franklin, R. E., *Proc. Roy. Soc.,* London, **A209**, 196 (1951).
[4] Dubinin, M. M., in: *Chemistry and Physics of Carbon,* P. L. Walker Jr. (Ed.), vol. 2, M. Dekker, New York 1966.
[5] Dubinin, M. M., *Izv. Akad. Nauk SSSR, Ser. Khim.* 1961 (1979).
[6] Kadlec, O., Choma, J., Jankowska, H. et al., *Collect. Czechoslov. Chem. Comm.* **49**, 2721 (1984).
[7] Smišek, M., Černý, S., *Active Carbon,* Elsevier, Amsterdam-London-New York 1970.
[8] Korver, J. A., *Chem. Weekbl.* **46**, 301 (1950).
[9] Dubinin, M. M., Zaverina, J. D., *Zh. Fiz. Khim.* **23**, 57 (1949); *Izv. Akad. Nauk SSSR*, O. Kh. N., 201 (1961).
[10] Jankowska, H., Świątkowski, A., Jarmoluk, A. et al., *Przem. Chem.* **59**, 563 (1980).
[11] Dziubalski, R., Korta, A., Smarzowski, J., *Koks, Smoła, Gaz* **24**, 253 (1979).
[12] Tarkovskaya, I. A., Okislennyi ugol', *Naukova Dumka,* Kiev 1981.
[13] Puri, B. R., in: *Chemistry and Physics of Carbon,* P. L. Walker Jr. (Ed), vol. 6, M. Dekker, New York 1970.
[14] Van der Plas, Th., in: *Physical and Chemical Aspects of Adsorbents and Catalysts,* Academic Press, London 1970.
[15] Mattson, J. S., Mark, H. B. Jr., *Activated Carbon,* M. Dekker, New York 1971.
[16] Cookson, J. T. Jr., in: *Carbon Adsorption Handbook,* Ann Arbor Science Publishers, Ann Arbor, Mich. 1978.
[17] Garten, V. A., Weiss, D. E., *Rev. Pure Appl. Chem.* **7**, 69 (1957).
[18] Boehm, H. P., *Advan. Catalysis* **16**, 179 (1966).
[19] Donnet, J. B., *Carbon* **6**, 161 (1968).
[20] Siedlewski, J., Śmigiel, W., *Wiadomości Chem.* **29**, 241 (1975).
[21] Siedlewski, J., Śmigiel, W., *Chem. Stosowana* **29**, 241 (1975).
[22] Garten, V. A., Weiss, D. E., *Australian J. Chem.* **10**, 309 (1957).
[23] Boehm, H. P., Voll, M., *Carbon* **8**, 227 (1970).
[24] Voll, M., Boehm, H. P., *ibid.* **9**, 481 (1971).

References

[25] Wolfrum, E. A., Dissertation, T. H., Aachen, *Berichte der Kernforschungsanlage Jülich*, No. 1194 (1975).
[26] Tarkovskaya, I. A., Strazhesko, D. N., Goba, W. E. et al., *Adsorbtsiya i Adsorbenty* **5**, 3 (1977).
[27] Boehm, H. P., Diehl, E., Heck, W. et al., *Angew. Chem.* **76**, 742 (1964).
[28] Puri, B. R., Singh, G., Sharma, L. R., *J. Indian Chem. Soc.* **34**, 357 (1957).
[29] Puri, B. R., Bansal, R. C., *Carbon* **1**, 457 (1964).
[30] Matsumura, Y., *J. Appl. Chem. Biotechnol.* **25**, 39 (1975).
[31] Weller, S., Young, T. F., *J. Am. Chem. Soc.* **70**, 4155 (1948).
[32] Vignaud, R., Brenet, J., *Compt. rend.* **257**, 3362 (1963).
[33] Vignaud, R., ibid., **262**, 952 (1966), *J. chim. phys.* **67**, 973 (1970).
[34] Vignaud, R., Dissertation, *Fac. des Sc. Univ. de Strasbourg*, 1968.
[35] Jankowska, H., Pietrzyk, S., Supruniuk, F., et al., *Biul. WAT* **23**, No. 9, 91 (1974).
[36] Pietrzyk, S., Jankowska, H., *ibid.* **23**, No. 9, 97 (1974).
[37] Gagnon, E. G., *J. Electrochem. Soc.* **122**, 521 (1975).
[38] Żółtowski, P., *J. Power Sources* **1**, 285 (1977).
[39] Ponomarenko, E. A., Frumkin, A. N., Burshtein, R. Kh., *Izv. Akad. Nauk SSSR, Ser. Khim.* 1549 (1963).
[40] Burshtein, R. Kh., Wilinskaya, W. S., Zagudaeva, N. M., et al., *Elektrokhimiya* **11**, 1882 (1975).
[41] Kastening, B., Schiel, W., Henschel, M., *J. Electroanal. Chem.* **191**, 311 (1985).
[42] Mioduska, M., Pietrzyk, S., Świątkowski, A., et al., *Biul. WAT*, **29**, No. 11, 119 (1980).
[43] Mioduska, M., Pietrzyk, S., Świątkowski, A., et al., *Biul. WAT* **28**, No. 9, 119 (1979).
[44] Puri, B. R., Bansal, R. C., *Carbon* **1**, 451 (1964).
[45] Coltharp, M. T., Hackerman, N., *J. Phys. Chem.* **72**, 1171 (1968).
[46] Barton, S. S., Gillespie, D., Harrison, B. H., *Carbon* **11**, 649 (1973).
[47] Voll, M., Boehm, H. P., *Carbon* **8**, 741 (1970).
[48] Rivin, D., *Rubber Chem. Technol.* **44**, 307 (1971).
[49] Vignaud, R., *Compt. rend.* **262**, 1609 (1966).
[50] Siedlewski, J. Śmigiel, W., *Chem. Stosowana* **21**, 175 (1977).
[51] Pietrzyk, S., Żmijewski, T., Mioduska, M., *Przem. Chem.* **58**, 496 (1979).
[52] Puri, B. R., *Carbon* **4**, 391 (1966).
[53] Puri, B. R., Singh, D. D., Sharma, L. R., *J. Phys. Chem.* **62**, 756 (1958).
[54] Barton, S. S., Dacey, J. R., Evans, M. J. B., *Colloid Polym. Sci.* **260**, 726(1982).
[55] Barton, S. S., *ibid.*, **264**, 176 (1986).
[56] Hallum, J. V., Drushel, H. V., *J. Phys. Chem.* **62**, 110 (1958).
[57] Jones, J. F., Kaye, R. C., *J. Electroanal. Chem.* **20**, 213 (1969).
[58] Bykov, V. T., Glushchenko, V. J., Ivashchenko, L. I., *Zh. Fiz. Khim.* **45**, 2884 (1971).
[59] Glushchenko, V. J. et al., *Zh. Fiz. Khim.* **46**, 187 (1972).
[60] Ivashchenko, L. I., Glushchenko, V. J., *Adsorbtsiya i Adsorbenty* **2**, 4 (1974).
[61] Darlewski, W., Rubaszkiewicz, J., *Biul. WAT* **25**, No. 7, 33 (1976).
[62] Kinoshita, K., Bett, J. A. S., *Carbon* **11**, 403 (1973).
[63] Świątkowski, A., Rubaszkiewicz, J., Bednarkiewicz, E., *J. Electroanal. Chem.* **239**, 91 (1988).
[64] Zagudaeva, N. M., *Elektrokhimiya* **22**, 1697 (1986).
[65] Jankowska, H., Neffe, S., *Przem. Chem.* **59**, 591 (1980).
[66] Jankowska, H., Gaczyński, R., Świątkowski, A., *Plaste u. Kautsch.* **27**, 256 (1980).
[67] Tomassi, W., Lewicki, W., *Przem. Chem.* **35**, 626 (1956).

[68] Tomassi, W., *ibid.* **38**, 77 (1959).
[69] Buława, J., *ibid.* **42**, 473 (1963).
[70] Tomassi, W., Ufnalski, W., *Biul. WAT* **23**, No. 5, 107 (1974).
[71] Świątkowski, A., Wieczorek, M., *Chem. Anal.* **29**, 49 (1984).
[72] Jankowska, H., Świątkowski, A., Wieczorek, M., *Chem. Stosowana* **25**, 565 (1981).
[73] Garten, V. A., Weiss, D. E., Willis, J. B., *Australian J. Chem.* **10**, 295 (1957).
[74] Friedel, R. A., Hofer, L. J. E., *J. Phys. Chem.* **74**, 2921 (1970).
[75] Ishizaki, C., Marti, I., *Carbon* **19**, 409 (1981).
[76] Mattson, J. S., Mark, H. B. Jr., *J. Colloid Interface Sci.* **31**, 131 (1969).
[77] Mattson, J. S., Mark, H. B. Jr., Weber, W. J. Jr., *Anal. Chem.* **41**, 355 (1969).
[78] Mattson, J. S., Lee, L., Mark, H. B. Jr., *et al.*, *J. Colloid Interface Sci.*, **33**, 284 (1970).
[79] Zawadzki, J., *Carbon* **16**, 491 (1978).
[80] Zawadzki, J., *ibid.* **19**, 19 (1981).
[81] Low, M. J. D., Morterra, C., *ibid.* **21**, 275 (1983).
[82] Glass, A. S., Low, M. J. D., *Spectrosc. Lett.* **19**, 397 (1986).
[83] Harker, H., Jackson, C., Wynne-Jones, W. F. K., *Proc. Roy. Soc. London* **A262**, 328 (1961).
[84] Ketov, A. N., Shenfeld, B. E., *Zh. Fiz. Khim.* **42**, 2104 (1968).
[85] Grishina, A. D., Semionov, A. P., *Electrokhimiya* **9**, 719 (1973).
[86] Dubinin, M. M., Zaverina, J. D., *Izv. Akad. Nauk SSSR*, O. Kh. N. 1038 (1956).
[87] Świątkowski, A., *Biul. WAT* **31**, No. 6, 81 (1982).
[88] Dubinin, M. M., Zaverina, E. D., Serpinskii, V. V., *J. Chem. Soc.* 1760 (1955).
[89] Dubinin, M. M., Serpinskii, V. V., *Dokl. Akad. Nauk SSSR* **258**, 1151 (1981).
[90] Jankowska, H., Choma, J., Paściak, T., *Wiadomości Chem.* **37**, 215 (1983).
[91] Evans, M. J. B., *Carbon* **25**, 81 (1987).
[92] Zawadzki, J., *Polish J. Chem.* **54**, 979 (1980).
[93] Jankowska, H., Lasoń, M., Polito, E., *et al.*, *Biul. WAT* **28**, No. 12, 73 (1979).
[94] Puri, B. R., Kumar, S., Sandle, N. K., *Indian J. Chem.* **1**, 418 (1963).
[95] Jankowska, H., Świątkowski, A., Ościk, J., *et al.*, *Carbon*, **21**, 117 (1983).
[96] Frumkin, A., *Kolloid Z.* **51**, 123 (1930).
[97] Kuchinsky, E., Burshtein, R., Frumkin, A., *Acta Physicochim.* USSR **12**, 795 (1940); *Zh. Fiz. Khim.* **14**, 441 (1940).
[98] Burshtein, R., Frumkin, A., *Dokl. Akad. Nauk SSSR* **32**, 327 (1941).
[99] Frumkin, A. N., Ponomarenko, E. A., Burshtein, R. Kh., *Dokl. Akad. Nauk SSSR* **149**, 1123 (1963).
[100] Steenberg, B., Thesis, Uppsala, Stockholm University, 1944.
[101] Ivanova, L. S., Svintsova, L. G., *Ukr. Khim. Zh.* **26**, 58 (1960).
[102] Ivanova, L. S., Strazhesko, D. N., *Adsorptsiya i Adsorbenty* **1**, 21 (1972).
[103] Jankowska, H., Świątkowski, A., Brzezicki, E., *Ukr. Khim. Zh.* **48**, 118 (1982).
[104] Lorenz, W., Reinhold, J., Siegemund, R. *et al.*, *Polish J. Chem.* **52**, 2431 (1978).
[105] Świątkowski, A., *Biul. WAT* **31**, No. 6, 93 (1982).
[106] Lavrinenko-Ometsinskaya, J. D., Strelko, V. V., Jankowska, H., *et. al.*, *Polish J. Chem.*, **63**, 197 (1989).
[107] Jankowska, H., Neffe, S., Świątkowski, A., *Electrochim. Acta* **26**, 1861 (1981).
[108] Jankowska, H., Starostin, L., *Przem. Chem.* **62**, 440 (1983).

CHAPTER 4
Models of Adsorption and their Corresponding Isotherms

4.1 GENERAL DESCRIPTION OF THE PHENOMENA

In the equilibrium established between the gas or vapour and the solid surface, the concentration of gas on the surface is usually greater than that in the gas phase irrespective of the natures of the gas and surface. The process in which an excess of the substance is accumulated on the surface of a solid is known as adsorption. The atoms on the surface of any solid experience the action of uncompensated attractive forces oriented perpendicular to the plane of the surface. This situation is partly restored on the adsorption of gas molecules. Solids with extensively developed surfaces that play the essential role in the process of adsorption are referred to as adsorbents. The real surface of the adsorbent per unit mass (usually 1 g) that participates in adsorption is called the specific surface area.

Experimental data show that, regardless of how the adsorption phenomena are described, the equilibrium distribution of the adsorbate molecules between the surface and the gas phase depends on the pressure of the gas undergoing adsorption and on the temperature:

$$a = f(p, T), \tag{4.1}$$

where a is the quantity of the adsorbed substance per unit surface area or mass of the adsorbent, p is the pressure of adsorbate under equilibrium conditions, and T is the temperature.

The adsorption equilibrium can be approached in three ways:

1. At constant temperature the equilibrium can be described by the adsorption isotherm:

$$a = f(p)_T. \tag{4.2}$$

2. At constant pressure, the equilibrium can be described by the adsorption isobar:

$$a = f(T)_p. \tag{4.3}$$

3. If the quantity a of the adsorbed substance is constant, the equilibrium can be described by the adsorption isostere:

$$p = f(T)_a. \tag{4.4}$$

In practice, experimental data on physical adsorption are usually presented in the form of adsorption isotherms. This is so because investigation of the adsorption process at constant temperature is most convenient, and theoretical analysis of the adsorption data as regards proposed models employs isotherms and not isobars or isosteres. It should be noted, however, that sometimes the adsorption isobars are very useful for determining the adsorption mechanism in a particular system and also for distinguishing between physical adsorption and chemisorption, determination of the temperature of desorption of a chemisorbed substance, etc.

To understand better the essence and mechanism of the adsorption process, one must become familiar with the commonest theories of adsorption and relationships describing the adsorption isotherms.

4.2 HENRY'S LAW

If we represent the adsorption equilibrium by the scheme:

molecule in the gas phase ⇌ adsorption complex (gas molecule on the adsorbent surface)

and assume in the case of a homogeneous surface that the concentration of the adsorbate in the surface layer is constant over the whole adsorbent surface, then the adsorption equilibrium is described by the equation:

$$\frac{c_s f_s}{cf} = K, \tag{4.5}$$

where c_s and c are the molar concentrations of the adsorbate in the surface layer and in the gas phase respectively, f_s and f are the activity coefficients for the respective phases, and K is the equilibrium constant (a function only of temperature).

At low concentrations in the gas phase ($f \approx 1$) and small values of c_s ($f_s \approx 1$), the concentration of adsorbed substance in the surface layer is:

$$c_s = Kc. \tag{4.6}$$

Since for ideal gases $c = p/RT$,

$$c_s = \frac{K}{RT} p. \tag{4.7}$$

The total amount a of the adsorbed substance in a volume v_s of the adsorption layer of 1 g of adsorbent is:

$$a = c_s v_s = \frac{K v_s}{RT} p. \tag{4.8}$$

For a given adsorption system, v_s and K are constant at constant temperature, so

$$a = K_{ap} p. \tag{4.9}$$

Equation (4.9) is the final form of the simplest equation of the adsorption isotherm called Henry's law.

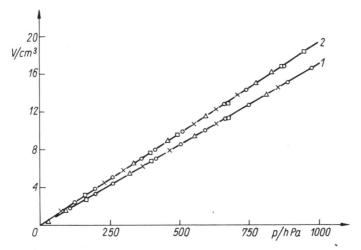

Fig. 4.1 Adsorption isotherms of 1 — oxygen, 2 — nitrogen on silica gel at 20°C (after Lambert and Peel [3]).

At low pressure p of the gas, the amount adsorbed a is proportional to the pressure in the bulk phase. It has been proved experimentally [1–4] that at room temperature, when the gas pressure is smaller than atmospheric pressure, the adsorption of nitrogen and oxygen on active carbon and silica gel is in agreement with Henry's law (Fig. 4.1).

Henry's isotherm can be theoretically derived if we assume that both the gas and the adsorbed phase are sufficiently dilute to consider them to be ideal. The general isotherm for this case, derived statistically by Wilkins [5], has the form:

$$a = p \frac{\alpha d}{RT} \frac{w_s}{w} \exp \frac{\varepsilon}{RT}, \qquad (4.10)$$

where a is the amount adsorbed, p is the pressure of adsorbate at equilibrium, α is that part of the surface area of the solid at which adsorption can take place, w and w_s are numbers the ratio of which is equal to that of the internal energies of the molecules in the gas and adsorbed phases, and ε is the adsorption potential assumed to be constant.

Thus the simplest adsorption isotherm, in which the amount adsorbed increases linearly with the equilibrium pressure of the gas, is called Henry's law by analogy with the isotherm for the solubility of gases in liquids.

4.3 LANGMUIR'S ISOTHERM

The origins of contemporary adsorption theory are associated with the name of Langmuir [5], the American researcher. To make possible a unique definition of the basic assumptions of Langmuir's isotherm, we will cite Fowler's formulation [6] which seems the most relevant in this case: "The necessary and sufficient assumptions of this theory are limited to: that the atoms (or molecules) of the gas are adsorbed at given active sites of the adsorbent surface so that every such site can bind one and only one adsorbed atom, and that the energy of state of every adsorbed atom is independent of whether other atoms are adsorbed or not in the neighbourhood of the considered site".

Thus the Langmuir theory deals with an ideally localized monomolecular layer with no intermolecular interactions.

In the original kinetic approach it is assumed that the adsorbed layer is in dynamic equilibrium with the gas phase. The number of molecules hitting unit surface area in unit time is, according to the kinetic theory:

$$n = \frac{p}{(2\pi mkT)^{1/2}}. \qquad (4.11)$$

The molecules impinging on the already occupied sites return immediately to the gas phase as if they were scattered. From among the molecules impinging upon an unoccupied active site, a definite fraction

α is retained by the surface forces for a finite time (this fraction of molecules is regarded as adsorbed) while the remaining fraction is scattered. If the fraction of occupied active sites is θ then the rate v_{ad} of conversion of the molecules to the adsorbed state is:

$$v_{ad} = \alpha n(1-\theta). \tag{4.12}$$

The rate v_{de} at which these molecules desorb is:

$$v_{de} = v\theta, \tag{4.13}$$

where v is the rate of desorption from the completely covered surface.

At equilibrium the number of molecules existing in the adsorbed state is constant and the rates of both processes are equal

$$v_{ad} = v_{de},$$

hence:

$$\alpha n(1-\theta) = v\theta. \tag{4.14}$$

Substituting the value for n from Eq. (4.11) into Eq. (4.14) and putting:

$$b = \frac{\alpha}{v(2\pi mkT)^{1/2}} \tag{4.15}$$

we obtain:

$$bp = \frac{\theta}{1-\theta} \tag{4.16}$$

and after transformation:

$$\theta = \frac{bp}{1+bp}. \tag{4.17}$$

The expression obtained is known as the Langmuir adsorption isotherm.

Denoting by V and V_m the volumes adsorbed at pressure p and at infinite pressure (when all active sites are occupied) we can write:

$$\theta = V/V_m. \tag{4.18}$$

Equation (4.17) assumes now the form:

$$V = \frac{V_m bp}{1+bp} \tag{4.19}$$

so V tends asymptotically to V_m as p tends to infinity.

At very low pressures, bp is much smaller than unity so to a first approximation Eq. (4.19) reduces to:

$$V = V_m bp. \qquad (4.20)$$

Equation (4.20) is an analytical representation of Henry's law.

For low pressures, the amount adsorbed is proportional to the first power of the pressure (4.20), while at high pressures it becomes pressure-independent. It can therefore be expected that at medium pressures the proportionality will be expressed by a fractional exponent $1/n$ which will tend to unity or to zero as p decreases or increases, respectively. This can be expressed by the general adsorption equation:

$$V = bp^{1/n} \qquad (4.21)$$

This is known as the Freundlich or Boedeker–Freundlich isotherm [6, 7] which is followed only at medium pressures.

To determine the values of the constants V_m and b in the Langmuir isotherm (4.19), this equation must be transformed to the linear form

$$\frac{p}{V} = \frac{1}{bV_m} + \frac{p}{V_m}, \qquad (4.22)$$

convenient for analysis.

Figure 4.2 shows the linear Langmuir isotherms for the adsorption of gases on active carbon plotted according to Equation (4.22) [8].

It should be noted, however, that the fact that a given isotherm satisfies the mathematical formalism of the Langmuir theory is not equivalent to saying that it meets the requirements of the model of an ideally localized monolayer, even if reasonable values of the coefficients b and V_m are obtained. This is because the non-homogeneity of the surface (non-equivalence of the adsorption sites, variability of the adsorbate–adsorbent interactions) may be compensated by adsorbate–adsorbent interactions and thus lead to a purely formal adherence to the Langmuir isotherm. Thus when determining the adsorption isotherm it is also necessary to determine the differential heat of adsorption Q which, according to the Langmuir theory, should be independent of the degree of coverage θ of the surface (in fact, the heat of adsorption almost always decreases with increase of θ).

The utilization of the Langmuir isotherm requires knowledge of the parameters V_m and b. For example, if we know V_m, the specific surface area S of the adsorbent can be determined, provided the surface area occupied by a single molecule in the monolayer, i.e. the so-called settling

Sec. 4.4] The BET isotherm 113

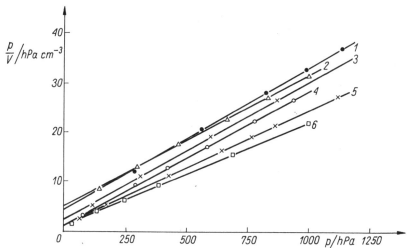

Fig. 4.2 The Langmuir adsorption isotherms of gases on active carbon: 1 — N_2 at 90 K, 2 — CO_2 at 195 K, 3 — CO at 90 K, 4 — N_2 at 77 K, 5 — Ar at 90 K, 6 — O_2 at 90 K (after Brunauer and Emmett [8]).

surface ω, is known. If we know the equilibrium constant of adsorption, $b = K$, the change of the standard free energy can be determined from the equation:

$$\Delta G^0 = \Delta H^0 - T\Delta S^0 = -RT \ln K'. \qquad (4.23)$$

The Langmuir theory relates to the ideal case when adsorption proceeds on a fully homogeneous surface. Nevertheless, the Langmuir model is still of great theoretical importance as it provides a good basis for succeeding, more complex models. This model finds particularly wide application in catalysis for exploring the kinetics of heterogeneous reactions.

The assumptions that the adsorption centres on the surface of a solid are homogeneous as regards their energies [9, 10] and that no interactions occur between the adsorbed molecules [11] are the weakest points of the Langmuir theory.

4.4 THE BRUNAUER, EMMETT AND TELLER (BET) ISOTHERM FOR MULTILAYER ADSORPTION

Langmuir tried as early as 1918 to generalize the concept of an ideally localized monolayer so that it would include the formation of multimolecular layers of adsorption. However, it was only in 1938 that an equation suitable for practical use was derived. The theory of multilayer adsorption developed by Brunauer, Emmett and Teller [12] and commonly known as the BET theory has played a dominant role in adsorption studies since it enables the specific surface area of the adsorbent and the approximate values of heats of adsorption to be determined.

The basic assumption of the BET theory is that the Langmuir isotherm can be applied to every adsorption layer. This generalization of the concept of an ideally localized monolayer is based on the assumption that every layer of adsorbed molecules constitutes a base for the adsorption of the molecules of the second layer which in turn acts as a base for the third layer, etc. Thus the concept of localization is maintained in all layers. Here also forces of interaction between the adsorbed molecules are neglected. With increasing pressure of the adsorbed vapours, i.e. as p approaches the saturated vapour pressure p_0, the number of unoccupied sites on the adsorbent surface decreases. Single, double, triple, etc. adsorption complexes are formed. The BET adsorption isotherm may be derived in several ways. Here the kinetic method is presented.

Let $S_0, S_1, S_2, S_3, \ldots, S_i, \ldots$, denote the surface areas covered by 0, 1, 2, 3, $\ldots$, i-molecular layers of the adsorbate. At equilibrium, the rate of condensation on surface area S_0 equals the rate of evaporation from surface area S_1, hence we have:

$$a_1 p S_0 = b_1 S_1 \exp\left(-\frac{\Delta E_1}{RT}\right), \tag{4.24}$$

where ΔE_1 is the heat of adsorption in the final layer, and a_1, b_1 are constants. Analogously the equilibrium between the first and the second adsorption layers can be written in the form:

$$a_2 p S_1 = b_2 S_2 \exp\left(-\frac{\Delta E_2}{RT}\right). \tag{4.25}$$

The general equation of equilibrium between layer $i-1$ and layer i is:

$$a_i p S_{i-1} = b_i S_i \exp\left(-\frac{\Delta E_i}{RT}\right). \tag{4.26}$$

The total surface area of the solid can be calculated from the equation:

$$S = \sum_{i=0}^{\infty} S_i \tag{4.27}$$

Sec. 4.4] The BET isotherm

and the total amount adsorbed from the formula:

$$a = a_0 \sum_{i=0}^{\infty} iS_i, \quad (4.28)$$

where a_0 is the quantity of the adsorbed substance per unit surface area for complete coverage of that surface by a monolayer. Dividing Eq. (4.28) by Eq. (4.27) we obtain:

$$\frac{a}{a_0 S} = \frac{a}{a_m} = \frac{\sum_{i=0}^{\infty} iS_i}{\sum_{i=0}^{\infty} S_i}, \quad (4.29)$$

where a_m is the quantity of adsorbate necessary to cover unit mass of the adsorbent with a monomolecular layer, and has the same meaning as in the theory of the ideally localized monolayer.

The BET equation can be derived after making some simplifying assumptions:

$$\Delta E_2 = \Delta E_3 = \ldots = \Delta E_i = \Delta E_l, \quad (4.30)$$

where ΔE_l is the heat of condensation of the liquid, and

$$\frac{b_2}{a_2} = \frac{b_3}{a_3} = \ldots = \frac{b_i}{a_i}. \quad (4.31)$$

It is assumed therefore that the processes of evaporation and condensation in the second and further layers proceed in the same way as on the surface of a liquid.

Equation (4.24) can be transformed as follows:

$$S_1 = yS_0, \quad (4.32)$$

where:

$$y = \frac{a_1}{b_1} p \exp\left(\frac{\Delta E_1}{RT}\right) \quad (4.33)$$

By using Eq. (4.30) it is possible to transform in a similar way Eq. (4.25):

$$S_2 = xS_1, \quad (4.34)$$

where

$$x = \frac{a_2}{b_2} p \exp\left(\frac{\Delta E_1}{RT}\right) \quad (4.35)$$

and in the general case, by making use of relationship (4.31) we obtain:
$$S_i = xS_{i-1} = x^{i-1}S_1 = yx^{i-1}S_0 = cx^i S_0, \qquad (4.36)$$
where
$$c = \frac{y}{x} = \frac{a_1 b_2}{a_2 b_1} \exp\left(\frac{\Delta E_1 - \Delta E_l}{RT}\right). \qquad (4.37)$$
Substituting Eq. (4.36) into Eq. (4.29) we obtain:
$$\frac{a}{a_m} = \frac{cS_0 \sum_{i=0}^{\infty} i x^i}{S_0 \left\{1 + c \sum_{i=1}^{\infty} x^i\right\}}. \qquad (4.38)$$

The sum in the denominator is that of an infinite geometric progression:
$$\sum_{i=1}^{\infty} x^i = \frac{x}{1-x}, \qquad (4.39)$$
while the expression in the numerator can be transformed to:
$$\sum_{i=1}^{\infty} i x^i = x \frac{d}{dx} \sum_{i=1}^{\infty} x^i = \frac{x}{(1-x)^2}. \qquad (4.40)$$
Substituting expressions (4.39) and (4.40) into Eq. (4.38) we obtain:
$$\frac{a}{a_m} = \frac{cx}{(1-x)(1-x+cx)}. \qquad (4.41)$$

The quantity of the substance adsorbed on the free surface at saturation is infinitely great. Thus for $p = p_0$, x in Eq. (4.41) must be unity if a is to tend to infinity. Thus from Eq. (4.35) it follows that:
$$\frac{a_2}{b_2} p \exp\left(\frac{\Delta E_l}{RT}\right) = 1 \qquad (4.42)$$
when
$$x = p/p_0. \qquad (4.43)$$
Substituting Eq. (4.43) into Eq. (4.41) we obtain:
$$a = \frac{a_m c \dfrac{p}{p_0}}{\left(1 - \dfrac{p}{p_0}\right)\left[1 + (c-1)\dfrac{p}{p_0}\right]}. \qquad (4.44)$$

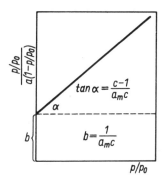

Fig. 4.3 Determination of the constants in the BET isotherm.

This equation is the so-called normal (for an infinite number of layers) BET adsorption isotherm. When Eq. (4.44) is transformed to the expression:

$$\frac{\frac{p}{p_0}}{a\left(1-\frac{p}{p_0}\right)} = \frac{1}{a_m c} + \frac{c-1}{a_m c} \cdot \frac{p}{p_0} \qquad (4.45)$$

one can easily see that the plot of $p/p_0/[a(1-p/p_0)]$ versus p/p_0 is a straight line whose slope is $(c-1)/(a_m c)$, and intercept is $1/(a_m c)$ (see Fig. 4.3). If we assume that

$$\frac{a_1 b_2}{a_2 b_1} \approx 1, \qquad (4.46)$$

then considering Eq. (4.37) we can write:

$$c = \exp\left(\frac{\Delta E_1 - \Delta E_l}{RT}\right). \qquad (4.47)$$

The term $\Delta E_1 - \Delta E_l$, i.e. the difference between the heats of adsorption in the first layer and of condensation, is known as the pure heat of adsorption. If we know the value of a_m, i.e. the quantity of adsorbate (mmol g^{-1}) covering the surface of the adsorbent with a monomolecular layer, we can calculate the specific surface area of the given adsorbent:

$$S_{\text{BET}} = a_m \cdot N \cdot \omega \qquad (4.48)$$

where N is Avogadro's number.

According to the classical BET procedure the specific surface area of adsorbents is measured by low-temperature (77 K) adsorption of nitrogen on the assumption that the settling surface area of one molecule is $\omega = 0.162$ nm^2.

In ref. [13] a numerical program is proposed for calculating the specific surface area of adsorbents by the BET method. The program enables rejection of doubtful measurements and calculation of the maximum relative error of approximation and is very useful for data processing.

If, due to inadequate space in the capillaries, adsorption at saturation is limited to n layers, then the BET theory leads us to the following isotherm [12, 14]:

$$a = \frac{a_m c x \left[1-(n+1)x^n + nx^{n+1}\right]}{(1-x)\left[1+(c-1)x - cx^{n+1}\right]} \tag{4.49}$$

where x, c and a_m have the same meaning as in Eq. (4.44).

Equation (4.49) is the most general form of the BET isotherm and it reduces to the Langmuir isotherm if $n = 1$ and to the simple form of the BET equation, i.e. to Eq. (4.44) if $n = \infty$. Some authors [15] claim that the concept of specific surface area loses its physical meaning in the case of microporous adsorbents. The BET theory is especially criticised as regards the theory of volume filling of micropores. According to this theory it is believed that application of methods based on multilayer adsorption (BET, de Boer's t-method) for determination of the specific surface area of adsorbents leads to results that have no physical justification [16, 17].

The present authors are of the opinion, based on experience and certain calculations [18], that the values of the specific surface area obtained by the BET method, even for typical microporous adsorbents, are realistic and have a physical meaning. The results obtained have also confirmed that it is useful to use the BET specific surface area to characterize the porous structure of adsorbents denoted as mesoporous or having a mixed micro- and mesoporous structure.

To find acceptable values of a_m, Eq. (4.44) should be used in the case of adsorption on meso- and macroporous adsorbents, and Eq. (4.49) in the case of microporous adsorbents. However, since the acceptability of the "specific surface area" parameter S_{BET} (m^2 g^{-1}) of the adsorbent depends on the adequacy both of the value found for the monolayer capacity a_m (mmol g^{-1}) and the assumed settling surface ω (nm^2 per molecule) of the adsorbate molecule, one should give values for a_m and ω in addition to S_{BET}.

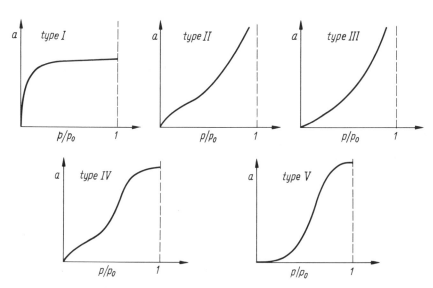

Fig. 4.4 The main types of isotherms of adsorption of gases and vapours (after Brunauer et. al. [14]).

It is well-known that the most typical deviations obtained when using the BET equation take the form of the predicted adsorption at low pressures being small and at high pressures being too high. Where p/p_0 lies between 0.05–0.3 the equation gives good agreement.

The adsorption isotherms given in the literature range from monotonic to highly complex. Brunauer *et al.* [14] claim that for systems kept at temperatures below the critical point of the gas, five principal shapes of the adsorption isotherms are possible (see Fig. 4.4).

Type I corresponds to the Langmuir isotherm: it tends monotonically to the limiting adsorption associated with a complete monolayer.

Type II corresponds to multilayer physical adsorption. This type of isotherm is obtained for the adsorption of benzene on graphitized carbon black.

Type III is relatively rare. Here the heat of adsorption is equal to or smaller (as regards its absolute value) than the heat of condensation of the pure adsorbate.

Isotherms of types IV and V differ from those of types II and III, respectively, in that in some sections they run parallel to the pressure axis. This is due to the generation in the adsorbent pores of a limited number of adsorption layers as a consequence of the widths of these pores.

4.5 THE HARKINS–JURA RELATIVE METHOD

Harkins and Jura [19, 20] derived on the basis of the model of formation of an adsorbate membrane on the surface of an adsorbent the following empirical equation of state for the so-called condensed film:

$$P = b - a\omega, \qquad (4.50)$$

where P is the two-dimensional surface pressure, ω is the mean surface occupied by one adsorbed molecule, and a and b are constants.

Using the Gibbs equation of adsorption we can transform Eq. (4.50) to a more convenient adsorption isotherm equation:

$$\log \frac{p}{p_0} = B - \frac{A}{V^2} \qquad (4.51)$$

known as the Harkins–Jura equation (H–J equation), where A and B are constants and V is the volume of the adsorbed vapour. Plots of (p/p_0) versus $1/V^2$ for several adsorbents of different specific surface areas are given in Fig. 4.5.

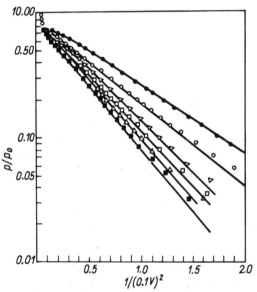

Fig. 4.5 Isotherms of adsorption of nitrogen at 77 K for various adsorbents plotted according to the H–J equation; the specific surfaces of the adsorbents are in m² g⁻¹: ● — 321, ○ — 365, ▽ — 395, □ — 401, △ — 438, ■ — 455 (after Jura and Harkins [19]).

Table 4.1

Comparison of the specific surface areas of adsorbents as determined by the H–J and BET methods (after Young and Crowell [21])

Adsorbent	Number of samples	S_{H-J}/S_{BET}		
		lowest	highest	mean
Carbon black	19	0.73	1.66	1.09
Inorganic pigments	13	0.92	1.19	1.01
Inorganic substances	8	0.85	1.11	1.04
Organic pigments	7	0.85	1.06	0.93
Graphites	4	0.89	1.05	0.97
Metals	3	0.73	0.93	0.84
Gels	3	0.95	0.98	0.97

The specific surface area of the adsorbent is related to the slope of the line of the H–J equation as follows:

$$S_{HJ} = kA^{1/2}, \qquad (4.52)$$

where k is a constant determined by an independent method for estimating the surface area. In consequence this method is often called the H–J relative method: the method is empirical. In the transformation of Eq. (4.50) into Eq. (4.51), thermodynamic relationships have been used but the final equation remains empirical just as the initial one. The range over which the H–J equation is consistent with experiment is usually greater than that of the BET equation.

In Table 4.1 the specific surface areas of various adsorbents as determined by the BET and H–J methods are compared. The last three columns provide the lowest, the highest and the mean values of the ratio S_{HJ}/S_{BET} for each group of adsorbents. The mean value of this ratio for all 57 samples is 1.02.

4.6 CAPILLARY CONDENSATION OF VAPOURS AND EVAPORATION OF THE CONDENSATE

Capillary condensation and capillary evaporation in the pores of adsorbents and catalysts are often utilized to describe quantitatively the volume and surface area distributions of the pores. On the surface of the pores, with the exception of micropores, adsorption proceeds at low relative pressures as on a non-porous material. Multilayer adsorption proceeds on the walls of macropores at medium and high relative pressures. By contrast condensation proceeds at medium pressures in mesopores. Capillary condensation in mesopores (intermediate pores) refers to the phenomenon of condensation of the adsorbate in the pores before the saturation vapour pressure is reached over the flat surface. The phenomenon of capillary condensation is due to the concavity of the surface layers of adsorbate covering the walls of the capillaries and hence the vapour pressure over them is lower than that over the flat surface. Thus condensation proceeds at pressures lower than that of the saturated vapour of the adsorbate.

The relationship between the pressure p of the saturated vapour over a curved surface of radius r_k and the pressure of the saturated vapour over the flat surface ($r_k = \infty$) has been given by Thomson (Kelvin):

$$\ln \frac{p}{p_0} = \frac{2\sigma V_m}{r_k RT}, \tag{4.53}$$

where σ is the surface tension and V_m the molar volume of the liquid, R is the universal gas constant, and T is the temperature.

The equation relating the pressure of the saturated vapour over a cylindrical liquid meniscus to the radius r_c of the curvature of that meniscus has the form [22, 23]:

$$\left(\ln \frac{p}{p_0} = -\frac{\sigma V_m}{r_c RT}\right). \tag{4.54}$$

It follows from Eqs. (4.53) and (4.54) that in the case of cylindrical, two-sided open pores the capillaries of a given diameter (the radii of curvature of the spherical and cylindrical menisci being equal, $r_k = r_c$) are filled in the process of adsorption with the condensed adsorbate (according to Eq. 4.53) under increased pressure and they are emptied during desorption (according to Eq. 4.54) under decreased pressure. Such a mechanism of filling and emptying of these pores is due to the irreversibility of the adsorption-desorption process as evident from the hysteresis of the isotherm [24].

The adsorption and desorption isotherms are used to describe quantitatively the distribution of pore volumes as a function of their radius. The theoretical descriptions used for this purpose assume a regular geometry for the pores. Thus the volume and surface area distribution functions determined relate to an equivalent model adsorbent, usually assuming a cylindrically-shaped pore, and no the real adsorbent [25]. It follows from the initial conditions and assumptions that the real adsorbent and the corresponding model are characterized by identical isotherms for capillary condensation and capillary evaporation. In contrast to the convex surface of aerosol droplets, i.e. spherical particles of different size, the formation of a concave surface of a liquid proceeds under different conditions, i.e. at the walls of a solid (adsorbent) and under definite wetting conditions. In general, the adsorption surface area, composed of the mesopore walls, affects both the thickness of the adsorption layer and the curvature of the meniscus at equilibrium. The relevant theory is due to Deryagin [26, 27], following whom de Boer and Broekhoff carried out approximate calculations of the effect of the surface area of the pore walls on the chemical potential of the adsorbed substance [28].

The thermodynamic theory of capillary evaporation of de Boer and Broekhoff leads to a system of two initial equations for the model adsorbent with cylindrical pores:

$$RT \ln(p/p_0) - F(t) = \frac{\sigma V_m}{r-t} \tag{4.55}$$

for $t = t_e$

$$r - t_e = \frac{2\sigma V_m}{RT \ln \frac{p_0}{p}} + \frac{2}{(r-t_e) RT \ln \frac{p_0}{p}} \int_{t_e} (r-t) F(t) \, dt, \tag{4.56}$$

where:

$$F(t) = 2.303 \, RT \frac{C}{t^2} - D(t) \tag{4.57}$$

is a function defined by the Harkins–Jura equation of the t-curve:

$$\log \frac{p_0}{p} = \frac{C}{t^2} - D(t). \tag{4.58}$$

An understanding of these quantities and their notation is aided by Fig. 4.6 which illustrates the axial section of the meniscus of the liquid in a cylindrical pore. In the formulae given above, σ is the surface tension of the liquid adsorbate, V_m is the molar volume of the liquid and t_e is the mean statistical thickness of the adsorption layer in the mesopores of radius r. In Eq. (4.58) giving the relationship between the thickness t of the adsorption layer and the relative equilibrium pressure p/p_0, C is a constant parameter while the correction term $D(t)$ is, depending on the

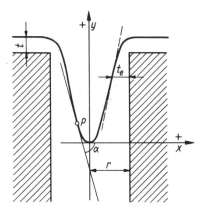

Fig. 4.6 Axial section of the meniscus in a cylindrical pore.

adsorption system, either a constant or a variable slightly dependent on t.

Equations (4.55) and (4.56) include three variables: p/p_0, t_e and r. When solving these equations with a computer we can, for example, determine the thickness t_e of the adsorption layer and radius r of the pores emptied due to capillary evaporation for a particular relative equilibrium pressure. The calculations made by Dubinin [29] have shown that the data obtained via the Broekhoff–Boer theory are incorrect for high relative equilibrium pressures. Attempts have been made to eliminate shortcomings of the theory by seeking an alternative version of the t-curve [29]. The analysis carried out showed it is worthwhile eliminating the term $D(t)$ from Eq. (4.58). In principle this is possible if a second parameter, in the form of an experimentally determined exponent of the quantity t, is introduced into the first term of the equation. Thus the revised t-curve equation takes the form of the Halsey equation [30]:

$$\ln \frac{p_0}{p} = \frac{K}{t^m}, \qquad (4.59)$$

where K and m are parameters.

The application of equation (4.59) in the Broekhoff and de Boer theory leads to simpler forms of the final equations. To obtain them, formula (4.59) should be substituted into formulae (4.55) and (4.56) together with a formula similar to Eq. (4.57) which has the form:

$$F(t) = RT \frac{K}{t^m}. \qquad (4.60)$$

After simple integration and algebraic transformations we finally obtain [29]:

$$\ln\frac{p_0}{p} - \frac{K}{t_e^m} - \frac{\sigma V_m}{RT(r-t_e)} = 0 \qquad (4.61)$$

$$r - t_e = \frac{2\sigma V_m}{RT \ln\frac{p_0}{p}} + \frac{2K}{(r-t_e)\ln\frac{p_0}{p}} \left[\frac{r^{2-m}}{(1-m)(2-m)} - \frac{rt_e^{1-m}}{1-m} + \frac{t_e^{2-m}}{2-m}\right]. \qquad (4.62)$$

The practical use of these equations in describing the texture of adsorbents is that their solution yields the so-called input data tables. A computer allows us to calculate on the basis of these equations the thickness t_e of the adsorbed layer and the radius r of the emptied pores for particular relative equilibrium pressures p/p_0, i.e. the points of the capillary evaporation isotherm. Experiments have confirmed that such data tables, i.e. of thicknesses of adsorbed layers and radii of emptied mesopores, for a complete range of relative equilibrium pressures changing in small increments, e.g. 0.02, are useful.

Table 4.2 summarizes the input data for the benzene/carbon system for the Kelvin, Broekhoff–de Boer, and Dubinin [31] methods which constitute a basis for further calculations.

Table 4.2
Input data for calculating the volume distribution of mesopores as a function of their radii for the active carbon–benzene system (temperature 20°C, r and t_e in nm) (after Dubinin and Kataeva [31])

p/p_0	Kelvin method		de Boer–Broekhoff method		Dubinin method	
	r	t	r	t_e	r	t_e
1	2	3	4	5	6	7
0.98	109.371	5.177	117.657	7.147	119.519	7.108
0.96	55.193	3.627	60.769	4.963	62.593	4.912
0.94	36.959	2.939	41.356	3.996	43.164	3.937
0.92	27.773	2.527	31.474	3.420	33.272	3.355
0.90	22.224	2.245	25.455	3.025	27.246	2.956
0.88	18.503	2.036	21.388	2.734	23.174	2.661
0.86	15.830	1.873	18.447	2.506	20.229	2.431
0.84	13.814	1.740	16.216	2.322	17.995	2.244
0.82	12.237	1.630	14.462	2.169	16.238	2.089
0.80	10.970	1.536	13.045	2.039	14.818	1.957
0.78	9.927	1.455	11.874	1.926	13.645	1.844

Table 4.2 (continued)

1	2	3	4	5	6	7
0.76	9.054	1.384	10.888	1.828	12.657	1.744
0.74	8.311	1.320	10.046	1.740	11.814	1.625
0.72	7.671	1.263	9.317	1.662	11.084	1.576
0.70	7.113	1.212	8.680	1.591	10.446	1.505
0.68	6.623	1.165	8.118	1.526	9.883	1.440
0.66	6.313	1.134	7.761	1.484	9.526	1.397
0.64	5.798	1.082	7.167	1.413	8.930	1.325
0.62	5.448	1.045	6.761	1.362	8.524	1.274
0.60	5.131	1.010	6.392	1.315	8.155	1.227
0.58	4.842	0.978	6.055	1.271	7.817	1.183
0.56	4.578	0.948	5.745	1.229	7.507	1.141
0.54	4.335	0.919	5.460	1.190	7.222	1.102
0.52	4.111	0.892	5.195	1.153	6.957	1.066
0.50	3.903	0.866	4.949	1.118	6.711	1.031
0.48	3.709	0.841	4.719	1.084	6.482	0.997
0.46	3.528	0.818	4.504	1.052	6.267	0.966
0.44	3.359	0.795	4.302	1.021	6.065	0.935
0.42	3.199	0.773	4.111	0.992	5.875	0.906
0.40	3.049	0.752	3.931	0.963	5.695	0.878
0.38	2.907	0.731	3.760	0.935	5.524	0.851
0.36	2.772	0.712	3.597	0.909	5.361	0.825
0.34	2.643	0.692	3.441	0.883	5.207	0.799
0.32	2.521	0.673	3.292	0.857	5.058	0.775
0.30	2.403	0.625	3.148	0.832	4.916	0.751
0.28	2.290	0.637	3.010	0.808	4.778	0.727
0.26	2.181	0.619	2.876	0.784	4.646	0.704
0.24	2.076	0.601	2.747	0.760	4.517	0.681
0.22	1.973	0.583	2.620	0.736	4.391	0.658
0.20	1.873	0.565	2.496	0.712	4.269	0.636
0.18	1.775	0.548	2.374	0.689	4.148	0.613
0.16	1.678	0.530	2.252	0.664	4.029	0.590
0.14	1.582	0.511	2.131	0.640	3.910	0.567
0.12			2.009	0.614	3.789	0.543
0.10			1.884	0.588	3.667	0.519
0.08			1.754	0.559	3.540	0.492
0.06			1.614	0.528	3.404	0.463

It should be noted that Jagiełło and Klinik [32], following similar lines, solved the Broekhoff–de Boer equations for argon (77 K) and carbon adsorbents, and summarized tables with the input data for the relevant calculations.

A number of methods are given for calculating the volume and surface area distributions of mesopores [33–38]. These methods are based on the experimental adsorption and desorption isotherms of vapours and gases. Klinik showed [39] that the parameters characterizing the mesopores of active carbons and silica gels obtained from adsorption-desorption isotherms by the Pierce, Dubinin (the so-called second variant), Crenston–Inkley, Barrett, Joyner–Hallenda, and Dollimore–Heal methods differ only slightly from each other (not more than by 5%). Usually the mean size of the emptied mesopores are determined from the quantity of adsorbate that leaves these pores at the assumed degree of desorption. This quantity is determined from the desorption branch of the isotherm. The condensed liquid leaves the pores under a relative pressure related to the pore radii, surface tension and molar volume of the adsorbate, as well as to temperature, by the given functions. Jankowska *et al.* [40] suggest the use of an algorithm for numerical analysis of the volume and surface area distributions of mesopores in adsorbents and catalysts. These authors used the Dollimore–Heal method [37] for developing the algorithm and the computer program which gives a good description of these distributions and avoids the shortcomings of the earlier methods. The Shull method [33] presents difficulties as regards the processing of experimental data for Maxwellian or Gaussian distribution functions and, as found in [34], these distributions are useless for some adsorbents. The Pierce method [35] involves inaccurate simplifying assumptions.

It follows from the theory of capillary evaporation of the mesopores that the amount ΔV of desorbing substance is the sum of the amount ΔV_c of substance leaving the pores due to capillary evaporation and the amount ΔV_m of the substance desorbing from the layer adsorbed on the walls of pores, hence:

$$\Delta V = \Delta V_c + \Delta V_m. \tag{4.63}$$

The value of ΔV_c enables determination of the actual volume ΔV_p of the pores from the formula:

$$\Delta V_p = \Delta V_c R, \tag{4.64}$$

where R is a coefficient accounting for the thickness of the adsorption layer remaining in the pores.

The approach of Dollimore and Heal [37] led to the following final formula:

$$\Delta V_{pn} = R_n (\Delta V_n - \Delta t_n \Sigma S_p + 2\pi t_n \Delta t_n \Sigma L_p) \tag{4.65}$$

which can be used to calculate the volumes of the pores for the nth step of desorption on the basis of the amount ΔV_n of the desorbing substance in the nth step, the change Δt_n in the thickness of the adsorbed layer on the pore walls in the nth step, and the thickness t_n of the layer in this step, account being taken of both the total surface area ΣS_p and the total length ΣL_p of the pores emptied in the previous steps.

The computer-analysis of data provides clear advantages as regards speed and accuracy.

4.7 POTENTIAL THEORY OF ADSORPTION

The Polanyi potential theory [41, 42], although developed some time ago, is still accepted as correct and is the subject of contemporary studies. Its thermodynamic approach and the fact that it does not impose any detailed physical model are the reasons for its retention. Its main assumptions are discussed below.

The forces retaining a molecule at the surface of a solid decrease as its distance from the surface increases. Thus an adsorbed multilayer should be treated as if it were in an adsorption potential field. A porous solid together with its adsorbed layer may be compared to the Earth with its atmosphere. The attractive forces, acting at a given point of the adsorbed layer, are characterized by the adsorption potential ε which, by definition, is equal to the work done by the adsorption forces during the passage of a molecule from the gas phase to this point:

$$\uparrow \varepsilon = RT \ln \frac{p_0}{p}. \tag{4.66}$$

The adsorption potential corresponds therefore to the work of compression of one mole of the vapour from the equilibrium pressure p to the saturated vapour pressure p_0. The section of the adsorbed layer corresponding to the adsorption potential theory, after Polanyi, is shown in Fig. 4.7. The dashed lines represent the equipotential planes. The volumes between the adsorbent and the equipotential planes, for which the adsorption potential is successively $\varepsilon_0, \varepsilon_1, \varepsilon_2, \varepsilon_3, \ldots, 0$, are $W_1, W_2, W_3, \ldots, W_0$, respectively. The last of these values, i.e. W_0, is the total (limiting) volume of the adsorption space. As the adsorption potential decreases from ε_0 at the surface of the adsorbent to zero in the furthest adsorption layer, the volume W increases from zero to W_0. The process of generation of the adsorption layer can be characterized by the relationship $\varepsilon = f(W)$ which is virtually the distribution function. The adsorption potential is by definition indepedent of temperature, and

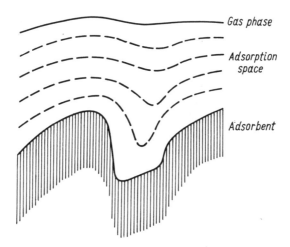

Fig. 4.7 Section of the adsorbed layer according to the potential adsorption theory (after Young and Crowell [21]).

hence the curve $\varepsilon = f(W)$ should be temperature-invariant for a given gas. Since this curve characterizes the given adsorbent-adsorbate system it is called the 'characteristic adsorption curve'. The temperature-independence of the adsorption potential can be described by the following equations:

$$\left(\frac{\partial \varepsilon}{\partial T}\right)_W = 0, \qquad (4.67)$$

$$\varepsilon = RT_1 \ln \frac{p_{0,1}}{p_1} = RT_2 \ln \frac{p_{0,2}}{p_2}. \qquad (4.68)$$

In contrast to other theories, Polanyi did not define an isotherm equation. At a certain stage of development of his theory, he used instead the characteristic adsorption curve presented in Fig. 4.8 by way of example for an active carbon.

Experimental verification of this theory depends on calculating the characteristic adsorption curve from one experimental isotherm and in determining on this basis the remaining isotherms. The basic isotherm used for determination of the characteristic adsorption curve should cover the whole range of ε values with great accuracy, as otherwise several isotherms would be required for exact determination of this curve.

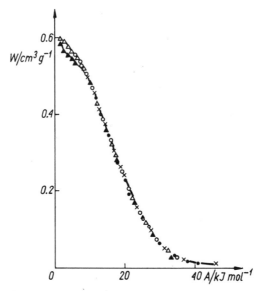

Fig. 4.8 Characteristic curve of benzene adsorption on active carbon SK:
▲ — 293 K, △ — 323 K, ○ — 353 K, × — 383 K, ● — 413 K (after Polanyi [43]).

When determining the characteristic adsorption curve, Polanyi [42] distinguished three separate cases:

(i) The temperature is much lower than the critical temperature of the adsorbate, in which case we are concerned with a liquid adsorption layer.

(ii) The temperature is marginally lower than the critical temperature; here the adsorption layer consists of liquid and compressed gas.

(iii) The temperature is higher than the critical temperature, when the adsorption layer consists of compressed gas.

Polanyi [43] and Berenyi [44, 45] also studied the relationships between the characteristic adsorption curves for various gases adsorbed on the same adsorbent. In the case of active carbon they found empirically that for any two gases the following equation is fulfilled:

$$\frac{\varepsilon_{01}}{\varepsilon_{02}} = \left(\frac{a_1}{a_2}\right)^{1/2}, \tag{4.69}$$

where a is the van der Waals constant.

Polanyi's potential theory of adsorption found a further, full development in the theory due to Dubinin et al. [46] of the volume filling of micropores which is discussed in the next chapter.

Attempts are made in more recent works to develop Polanyi's potential theory by applying a mathematical description of the adsorption equilibria that aims to relate the magnitude of adsorption a with the adsorption potential ε [47–48]. In ref. [48] this description is based on the assumption that the differential distribution of adsorption a as a function of the common logarithm of the reciprocal of the adsorption potential, $\log 1/\varepsilon$, can be a normal (Gaussian) distribution. This assumption also relates to the distribution of the volume W of the adsorbent pores as a function of the adsorption potential, since

$$a = \frac{W}{V_m}, \qquad (4.70)$$

where V_m is the molar volume of the liquid adsorbate.

The distribution of the adsorbed substance in the field of its adsorption forces is given by the relationship:

$$\frac{\partial a}{\partial \log 1/\varepsilon} = \frac{a_{max}}{\sigma\sqrt{2\pi}} \exp\left[-\frac{(\log 1/\varepsilon - \log 1/\varepsilon_m)^2}{2\sigma^2}\right], \qquad (4.71)$$

where a_{max} is the maximum adsorption, $\log 1/\varepsilon_m$ is the common logarithm of the reciprocal of the adsorption potential which corresponds to the maximum of the distribution function, and σ is the scattering (dispersion) characterizing the width of the distribution function.

The integral formula takes the form:

$$a = \int_{-\infty}^{\log 1/\varepsilon_a} \frac{a_{max}}{\sigma\sqrt{2\pi}} \exp\left[-\frac{(\log 1/\varepsilon - \log 1/\varepsilon_m)^2}{2\sigma^2}\right] d\log 1/\varepsilon, \qquad (4.72)$$

where $\log 1/\varepsilon_a$ is the upper limit of integration corresponding to the value of the adsorption sought.

The parameters of this relationship, a_{max}, σ and $\log 1/\varepsilon_m$ which are also the structural parameters of the adsorbent, have a specific and real physical meaning. The method proposed [47, 48] enables an accurate description of the adsorption process on active carbons for various adsorbates over a wide range of temperatures and relative pressures.

It is not easy to say which of the various theories of adsorption of vapours and gases is the best. Each has its advantages and disadvantages. It seems that the potential theory, developed as the theory of volume filling of micropores, is currently the most rational theory for describing adsorption phenomena occurring on microporous adsorbents.

4.8 ADSORPTION ON ACTIVE CARBONS WITH MICRO-, MESO-, AND MACROPORES

For active carbon we distinguish three types of pore: micro-, meso- and macropores which play different roles in the process of adsorption of vapours and gases. For typical widely-used active carbons, most adsorbate is adsorbed in the micropores. The classification of pores is usually based on their linear dimensions. According to the IUPAC classification of pores in adsorbents [49], which differs from that presented in Chapter 3, pores with radii below 2 nm are called micropores, those with radii lying between 2 and 50 nm are called mesopores, while those with radii greater than 50 nm are called macropores. The presence of micropores in the adsorbent changes substantially its adsorptive properties. The overlapping of the fields of adsorption forces generated by the opposite walls of the micropores leads to an substantial increase of the adsorption potential inside these pores [50]. As a result, adsorption in the micropores is much greater than on the surface of the mesopores [51]. Adsorption on the surface of macropores is usually negligible compared with that in micro- and mesopores [51].

Detailed specification of an active carbon requires: determination of the pore volume distribution function variation with the linear dimensions (with the micro-, meso- and macropore ranges being included), quantitative determination of the chemical functional groups present on the surface, estimation of the adsorption capacity and specific surface area of the particular types of pore, description of the structural and energetic heterogeneity of the pores, etc.

To characterize microporous active carbons it is necessary to know: (i) the estimated proportion of microporosity in the overall porosity of the adsorbent, and (ii) the values of the parameters describing the microporous structure. Most of the popular methods for estimating microporosity consist of comparing the isotherm determined for the given active carbon with the standard isotherm for a non-porous adsorbent which acts as the reference system. The t-method [52, 53] and the α_s-method [54, 55] are examples of such methods. In the case of the t-method the values of adsorption on the test adsorbent are given as a function of the statistical thickness t of the adsorption film on a non-porous standard adsorbent. The α_s-method differs from the t-method in that the adsorption values are given as a function of the reduced standard adsorption α_s which for the non-porous adsorbent is defined as the ratio of adsorption at the given relative pressure p/p_0 to adsorption at $p/p_0 = 0.4$. The pre-adsorption method [56–58] is another way of estimating microporosity, according to which large particles (e.g. of nonane) are used for filling (blocking) the micropores of the adsorbent. The use of adsorption methods for describing the microporosity of adsorbents has been also described by Sing [59], and Gregg and Sing [60]. All these studies have contributed significantly to the IUPAC report on the characteristics of porosity in solids [49].

The second important question related to the characterization of microporous active carbons is the determination of the parameters describing the microporous structure. In the case of homogeneous microporous active carbons, the Dubinin–Radushkevich (D–R) equation is usually applied for determining the volume and characteristic size of the micropores [61, 62]. For determining the parameters of micropores in heterogeneous microporous adsorbents, the Dubinin–Stoeckli [63] or Jaroniec–Choma [64, 65] equations are used. More details on this subject are given in Chapter 5.

Among the adsorption methods for estimating microporosity, the α_s-method proposed by Sing [54, 55, 59 and 60] is often used. This method can be applied for analysis of the experimental isotherms for one adsorbate (e.g. benzene) and its use involves fairly easy calculations of the reduced standard adsorption. A brief discussion of this method is presented later in this book. Adsorption a corresponding to the relative pressure p/p_0 is the sum of adsorption in the micropores (a_{mi}), adsorption on the surface of the mesopores (a_{me}), and that on the surface of the macropores (a_{ma}), all of which correspond to the relative pressure p/p_0, i.e.:

$$a = a_{mi} + a_{me} + a_{ma}. \tag{4.73}$$

In view of the very small specific surface area of the macropores (several m^2 per gram of active carbon) as compared with the specific surface area of micro- and mesopores, the adsorption in the macropores may be regarded as zero, i.e.

$$a_{ma} = 0 \tag{4.74}$$

and then the total adsorption becomes:

$$a = a_{mi} + a_{me}. \quad (4.75)$$

This equation may be written in a form taking account of the relative adsorption θ_{mi} and θ_{me} [65]:

$$a = a^0_{mi}\,\theta_{mi} + a^0_{me}\,\theta_{me}, \quad (4.76)$$

where

$$\theta_{mi} = \frac{a_{mi}}{a^0_{mi}} \quad (4.77)$$

and

$$\theta_{me} = \frac{a_{me}}{a^0_{me}}, \quad (4.78)$$

a^0_{mi} being the maximum adsorption in the micropores and a^0_{me} the capacity of the monolayer on the surface of the mesopores. For the standard experimental adsorption isotherm and a non-porous reference adsorbent we have:

$$\theta_s = \frac{a_{s(0.4)}}{a^0_s}\alpha_s \quad \text{where} \quad \alpha_s = \frac{a_s}{a_{s(0.4)}}, \quad (4.79)$$

θ_s is the coverage of the surface of the non-porous (reference) adsorbent, a_s is the value of the adsorption for a standard adsorbent at a relative pressure p/p_0, a^0_s is the capacity of the monolayer determined from the standard adsorption isotherm, and $a_{s(0.4)}$ is the value of adsorption for the standard adsorbent at a relative pressure $p/p_0 = 0.4$.

For large values of p/p_0 the micropores are completely filled which means that

$$\theta_{mi} = 1 \quad (4.80)$$

and the mechanism of formation of the multilayer on the surface of the mesopores is the same as that taking place during adsorption on the surface of the non-porous standard adsorbent [65]. Thus for the same values of the relative pressure p/p_0 the degrees of coverage are equal:

$$\theta_{me} = \theta_s. \quad (4.81)$$

By combining equations (4.76), (4.79) and (4.81) we obtain [65]:

$$a = a^0_{mi}\,\theta_{mi} + \varkappa\alpha_s, \quad (4.82)$$

where the constant $\varkappa$ is defined as:

$$\varkappa = \frac{a_{me}^0 \, a_{s(0.4)}}{a_s^0}. \tag{4.83}$$

According to Eq. (4.82) for low values of p/p_0, the variation of a as a function of α_s is not linear since the variable θ_{mi} is not proportional to α_s as the value of p/p_0 changes. However, for values of p/p_0 greater than the pressure necessary for filling the micropores, Eq. (4.80) holds, and then Eq. (4.82) reduces to the following linear relationship:

$$a = a_{mi}^0 + \varkappa \alpha_s. \tag{4.84}$$

This situation is illustrated in Fig. 4.9.

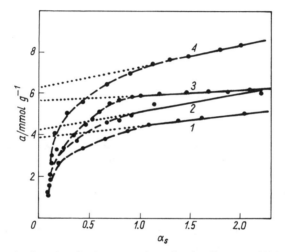

Fig. 4.9 A plot of α_s for benzene adsorption isotherms at 20°C on active carbon: 1 — AG-5, 2 — AC-12, 3 — T, and 4 — CWZ-3 (after Jaroniec et al. [65]).

To calculate the reduced standard adsorption α_s, the adsorption isotherm of benzene at 20°C measured for various types of carbon blacks is often used [66, 67].

Equation (4.84) allows us to calculate the volume V_{mi} of micropores and the specific surface area S_{me} of mesopores. The volume V_{mi} of micropores is proportional to a_{mi}^0:

$$V_{mi} = a_{mi}^0 \, V_m, \tag{4.85}$$

where V_m is the molar volume of the adsorbate.

The surface area S_{me} of mesopores is in turn proportional to the constant $\varkappa$:

$$S_{me} = \varkappa N \omega, \tag{4.86}$$

where N is Avogadro's number, and ω is the surface area occupied by one molecule of adsorbate in the monolayer.

To determine the real parameters of the microporous structure of the adsorbent it is necessary to account for adsorption occurring on the mesopore surface. This is done by subtracting the adsorption in the mesopores from the experimental isotherm of the overall adsorption. Then expression (4.75) assumes the form:

$$a_{mi} = a - a_{me}. \tag{4.87}$$

The adsorption in the mesopores may be written in the form:

$$a_{me} = \gamma S_{me}, \tag{4.88}$$

where γ (in mmol m^{-2}) is the adsorption of gas or vapour at a given temperature per unit surface area of the adsorbent, and S_{me} is the specific surface area of the mesopores in m^2 g^{-1}. The function γ may be determined from the experimental adsorption data obtained for a specially prepared non-porous carbon black. Based on the measurements of adsorption of benzene on carbon black 950, Dubinin [63] derived the following equation for the standard isotherm:

$$\gamma = \gamma^0 \exp\left(-\frac{A}{6.35}\right), \tag{4.89}$$

where: $\gamma^0 = 9.16 \times 10^{-3}$ mmol m^{-2}, and A is the differential molar work of adsorption given by the formula:

$$A = RT \ln(p_0/p), \tag{4.90}$$

where R is the universal gas constant, T is the absolute temperature and p/p_0 is the relative equilibrium pressure. Equation (4.89) may be applied over a range of relative pressures from 10^{-5} to 0.3.

To provide for adsorption in the mesopores it is necessary according to relationship (4.88), to know the specific surface area S_{me} of the mesopores as determined from adsorption data. Apart from the α_s-method discussed above, several other methods of determining S_{me} are described in the literature [37, 68–72].

One of these is the method described in refs. [68, 69] involving the integral relationship:

$$S' = \frac{1}{\sigma} \int_{a_0}^{a_s} RT \ln(p_0/p)\, da, \tag{4.91}$$

where S' is the surface area of the adsorption film formed on the surface of mesopores, assumed to be equal to the specific surface area of these mesopores ($S' = S_{me}$), σ is the surface tension of the adsorbate and a_s is the limiting values of the adsorption for the given system at equilibrium pressure $p = p_0$, and a_0 is the adsorption corresponding to the initial point of the hysteresis loop. The specific surface area of the mesopores is the arithmetic mean of the surface areas obtained for the adsorption and desorption branches of the adsorption isotherm hysteresis loop.

Another method of determining the surface areas S_{me} of the mesopores has been forwarded by Dollimore and Heal [37]. Although this method is primarily intended for determining the mesopore volume distribution as a function of their radii (see Eq. (4.65)), it may be also used for calculating S_{me} from the equation:

$$S_{me} = \sum_{i=1}^{n} \frac{2\Delta V_{pi}}{\bar{r}_i}, \tag{4.92}$$

where ΔV_{pi} is the decrease of the pore volume determined from the quantity of adsorbate desorbed in the ith step of desorption, $\bar{r}_i$ is the average radius of the pores emptied in the ith step of desorption, and n is the number of steps of desorption corresponding to the number of intervals into which the whole mesopore range has been divided.

An interesting method of determining the specific surface area of mesopores has been advanced by Dubinin and Kadlec [70]. By substituting Eq. (4.88) into Eq. (4.87) we obtain:

$$a = a_{mi} - \gamma S_{me}. \tag{4.93}$$

Dividing both sides of Eq. (4.93) by θ_{mi} determined by Eq. (4.77) we obtain:

$$\frac{a}{\theta_{mi}} = a_{mi}^0 + S_{me} \frac{\gamma}{\theta_{mi}}. \tag{4.94}$$

Equation (4.94), introduced by Dubinin and Kadlec, enables the determination of the specific surface area S_{me} of the mesopores, provided one can determine the degree of filling θ_{mi} of the adsorption space of the micropores. The specific surface area S_{me} of the mesopores is the slope of the straight line obtained by plotting the experimental data for a/θ_{mi}

versus γ/θ_{mi}. Dubinin and Polyakov [72] have suggested a method of estimating θ_{mi} for a given carbon adsorbent.

All methods of determining the specific surface S_{me} of the mesopores from adsorption data discussed in this chapter have been evaluated and compared in detail by Jaroniec and Choma [73]. These authors found that the different methods yield somewhat different values of the specific surface areas of the adsorbent mesopores and that these differences also affect the microporous structural parameters.

The observed differences show once more that the porous structure of active carbon is very complex and the methods used for its estimation are only approximate. A substantial advance in the theory of multilayer adsorption as applicable to mesopores offers some possibility of improving the description of adsorption on the surface of mesopores.

These various considerations allow us to conclude that the determination of the specific surface area of the mesopores is a complex problem but this should not be neglected when determining the parameters of the microporous structure of an adsorbent.

References

[1] Chaplin, R., *Phil. Mag.* **2**, 1198 (1926).
[2] Weissman, H., Neumann, W. Z., *Pflernahr. Dung.* **40**, 49 (1935).
[3] Lambert, B., Peel, D. H., *Proc. Roy. Soc.* (London) **A144**, 205 (1934).
[4] Emmett, P. H., Brunauer, S., Love, K., *Soil. Sci.* **45**, 47 (1938).
[5] Langmuir, I., *J. Am. Chem. Soc.* **40**, 1361 (1918).
[6] Boedeker, C., *J. Landw.* **7**, 48 (1895).
[7] Freundlich, H., *Colloid and Capillary*, Methuen, London 1926.
[8] Brunauer, S., Emmett, P. H., *J. Am. Chem. Soc.* **57**, 1754 (1935).
[9] Taylor, H. S., *Proc. Roy. Soc.* (London) **A108**, 105 (1925).
[10] Constable, F. H. *Proc. Roy. Soc.* (London) **A108**, 355 (1925).
[11] Laidler, K. J., *Catalysis, Chemisorption*, Reinhold Publishing Corp., New York 1954, pp. 75–118.
[12] Brunauer, S., Emmett, P. H., Teller, E., *J. Am. Chem. Soc.* **60**, 309 (1938).
[13] Hippe, Z., Dębska, B., Kerste, A., *Ćwiczenia z chemii fizycznej do obliczeń na EMC* (Exercises in Physical Chemistry with Programs for Computer Calculations), PWN, Warsaw 1979.
[14] Brunauer, S., Deming, L. S., Deming, W. E., Teller, E., *J. Am. Chem. Soc.* **62**, 1723 (1940).
[15] Bering, B. P., Dubinin, M. M., Serpinskii, V. V., *J. Colloid Interface Sci.* **39**, 185 (1972).
[16] Dubinin, M. M., *Izv. Akad. Nauk SSSR, Ser. Khim.* 9 (1981).
[17] Dubinin, M. M., Efremov, S. N., Kataeva, L. I., Ulin. V. I., *Izv. Akad. Nauk. SSSR, Ser. Khim.* 1711 (1981).
[18] Choma, J., *Polish J. Chem.* **57**, 507 (1983).
[19] Jura, G., Harkins, W. D., *J. Chem. Phys.*, **11**, 430 (1943).

[20] Jura, G., Harkins, W. D., *J. Am. Chem. Soc.* **68**, 1941 (1946).
[21] Young, D. M., Crowell, A. D., *Physical Adsorption of Gases*, Butterworths, London 1962.
[22] Cohan, L. H., *J. Am. Chem. Soc.* **60**, 443 (1938).
[23] Cohan, L. H., *J. Am. Chem. Soc.* **66**, 98 (1944).
[24] Korta, A., Studium sorpcji substancji polarnych i niepolarnych z par i z roztworów wodnych na sorbentach węglowych (Study of sorption of polar and non-polar substances from vapour and aqueous solutions on carbon sorbents), Academy of Mining and Metallurgy, Cracow 1967, Fasc. 13, *Górnictwo* (Mining), No. 189.
[25] Dubinin, M. M., *Zh. Fiz. Khim.* **30**, 1652 (1956).
[26] Deryagin, B. V., *Acta Physicochim.* **12**, 181 (1940).
[27] Deryagin, B. V., *J. Colloid Interface Sci.* **24**, 357 (1967).
[28] Broekhoff, J. C. P., de Boer, J. H., *J. Catalysis* **9**, 8 (1967).
[29] Dubinin, M. M., *Izv. Akad. Nauk SSSR, Ser. Khim.* 22 (1980).
[30] Halsey, G. D., *J. Chem. Phys.* **16**, 931 (1948).
[31] Dubinin, M. M., Kataeva, L. I., *Izv. Akad. Nauk SSSR, Ser. Khim.* 498 (1980).
[32] Jagiełło, J., Klinik, J., *Chem. Stosowana*, **25**, 573 (1981).
[33] Shull, C. G., *J. Am. Chem. Soc.* **70**, 1405 (1948).
[34] Barrett, E. P., Joyner, L. G., Halenda, P. P., *ibid.* **73**, 1951 (1951).
[35] Pierce, C., *J. Phys. Chem.* **57**, 149, 64 (1935).
[36] Cranston, R. W., Inkley, F. A., *Advan. Catalysis* **IX**, 143 (1957).
[37] Dollimore, D., Heal, G. R., *J. Appl. Chem.* **14**, 109 (1964).
[38] Jagiełło, J., Klinik, J., *Chem. Stosowana* **24**, 331 (1980).
[39] Klinik, J., Doctoral Thesis, Academy of Mining and Metallurgy, Cracow 1981.
[40] Jankowska, H., Choma, J., Szypuła, W., *Przem. Chem.* **60**, 390 (1981).
[41] Polanyi, M., *Vehr. deutsch. phys. Ges.* **16**, 1012 (1914).
[42] Polanyi, M., *ibid.* **18**, 55 (1916).
[43] Polanyi, M., *Z. Elektrochem.* **26**, 371 (1920).
[44] Berenyi, L., *Z. physik. Chem.* **94**, 628 (1920).
[45] Berenyi, L., *ibid.* **105**, 55 (1923).
[46] Dubinin, M. M., *Chemistry and Physics of Carbon* (Ed. Walker, P. J., Jr.) **2**, 51, M. Dekker, New York 1966.
[47] Choma, J., *Polish J. Chem.* **58**, 201 (1984).
[48] Choma, J., *Przem. Chem.* **62**, 634 (1983).
[49] Sing, K. S. W., Everett, D. H., Haul, R. A. W., Moscou, L. et al. *Pure Appl. Chem.* **57**, 603 (1985).
[50] Everett, D. H., Powl. J. C., *J. Chem. Soc. Faraday Trans. I*, **72**, 619 (1976).
[51] Dubinin, M. M., *Carbon* **21**, 359 (1983).
[52] Lippens, B. C., de Boer, J. H., *J. Catalysis*, **4**, 319 (1965).
[53] Sing, K. S. W., *Chem. Industry* **67**, 829 (1967).
[54] Sing, K. S. W., *Surface Area Determination* (Ed. Everett, D. H.), Butterworths, London 1970.
[55] Carott, P. J. M., Roberts, R. A., Sing, K. S. W., *Carbon* **25**, 59 (1987).
[56] Gregg, S. J., Langford, J. F., *Trans. Faraday Soc.* **65**, 1394 (1969).
[57] Rodriguez-Reinoso, F., Martin-Martinez., J. M., Molina-Sabio, M. et. al., *J. Colloid Interface Sci.* **106**, 315 (1985).
[58] Ali, S., McEnaney B., *ibid.* **107**, 355 (1985).
[59] Sing, K. S. W., *Berichte Bunsen-Gesellschaft Phys. Chem.* **79**, 724 (1975).

[60] Gregg, S. J., Sing, K. S. W., *Adsorption, Surface Area and Porosity*, 2nd ed., Academic Press, London 1982.
[61] Dubinin, M. M., *Progress in Surface and Membrane Science* (Ed. Cadenhead, D. A.) **9**, 1–70, Academic Press, New York 1975.
[62] Dubinin M. M., Characterization of Porous Solids (Eds. Gregg, S. J., Sing, K. S. W., Stoeckli, H. F.) 1–11, Society of Chemical Industry, London 1979.
[63] Dubinin, M. M., *Carbon* **23**, 373 (1982).
[64] Jaroniec, M., Choma, J., *Materials Chem. Phys.* **15**, 521 (1986).
[65] Jaroniec, M., Madey, R., Choma, J., et. al. *Carbon* **27**, 77 (1989).
[66] Dubinin, M. M., Izotova, T. I., Kadlec, O., et. al. *Izv. Akad. Nauk SSSR, Ser. Khim.* **75**, 1232 (1975).
[67] Choma, J., Jaroniec, M., *Przem. Chem.* **67**, 499 (1988).
[68] Kistler, S. S., Fischer, E. A., Freeman, J. H., *J. Am. Chem. Soc.* **65**, 1909 (1943).
[69] Kiselev, A. V., Mikos, N. N., *Zh. Fiz. Khimii* **22**, 1043 (1948).
[70] Dubinin, M. M., Kadlec, O., *Carbon* **13**, 263 (1975).
[71] Stoeckli, H. F., Kraehenbuehl, F., *Carbon* **22**, 297 (1984).
[72] Dubinin, M. M., Polakov, N. S., *Izv. Akad. Nauk SSSR, Ser. Khim.* **86**, 1932 (1986).
[73] Choma, J., Jaroniec, M., *Materials Chem. Phys.* **18**, 409 (1987).

CHAPTER 5
The Theory of Volume Filling of Micropores and its Developments

5.1. FOUNDATIONS OF THE THEORY

The theory of volume filling of micropores (TVFM) is based on the assumption that the characteristic adsorption equation expressing the distribution of the degree of filling, θ_{mi}, of the adsorption space, i.e. the volume of the micropores as a function of the differential molar work of adsorption A, is temperature-independent. This theory describes satisfactorily the shape of the adsorption isotherms where vapours adsorb strongly on microporous active carbon. In this case dispersion forces are considered to be the main component of the interaction between adsorbent and adsorbate. This theory, mainly elaborated by Dubinin et al. [1–6], has been further extended and improved.

Dubinin, in contrast to Polanyi who defines the differential molar work of adsorption A as the adsorption potential, defines A as the change in the Gibbs free energy:

$$A = -\Delta G \tag{5.1}$$

with

$$A = RT \ln \frac{p_0}{p}, \tag{5.2}$$

where p_0 is the saturated vapour pressure of the substance considered at temperature T and p is the equilibrium pressure.

The fundamental TVFM equation in its most general form may be presented as follows:

$$\theta_{mi} = f\left(\frac{A}{E}, n\right), \tag{5.3}$$

where E and n are the distribution function parameters and θ_{mi} is the degree of filling of the adsorption volume defined as:

$$\theta_{mi} = \frac{a_{mi}}{a_{mi}^0}, \qquad (5.4)$$

where a_{mi} is the adsorption at temperature T and pressure p at the equilibrium state and a_{mi}^0 is the limiting adsorption value that corresponds to complete filling of the entire adsorption volume.

Analysis of experimental data made it possible to present relation (5.3) in the following form suitable for practical application:

$$\theta_{mi} = \exp\left[-\left(\frac{A}{E}\right)^2\right], \qquad (5.5)$$

where E is a distribution function parameter known as the characteristic energy of adsorption.

If E_0 is the characteristic energy of adsorption of a standard vapour (usually of benzene on active carbon), then the characteristic adsorption energy E of another vapour (or gas) is given by the expression:

$$E = \beta E_0, \qquad (5.6)$$

where β is a coefficient called by different authors the similarity or affinity coefficient. This coefficient is fairly well approximated by the ratio of the parachors i.e. P is that of the test substance and P_0 that of the standard substance calculated additively according to Sugden from the atomic and bond components [7]:

$$\beta \approx \frac{P}{P_0}. \qquad (5.7)$$

The limiting adsorption a_{mi}^0 is temperature dependent due to the thermal expansion of the adsorbed substance (the temperature variation of the adsorption space limiting volume W_{mi}^0 can be neglected)[†]. If by ϱ^* we denote the density and by V_m the molar volume of the adsorbate of molecular mass M at the limiting filling of the micropores, then:

$$a_{mi}^0 = W_{mi}^0 \frac{1}{M} \varrho^* = \frac{W_{mi}^0}{V_m}. \qquad (5.8)$$

[†] Note that in the classical D–R equation the limiting adsorption was denoted as a_0 and the limiting volume of the adsorption space as W_0.

Sec. 5.1] Foundations of the theory

On substituting equations (5.2), (5.4), (5.6) and (5.8) into the adsorption equation (5.5), we obtain the equation for the adsorption isotherm, known as the Dubinin–Radushkevich (D–R) equation, in the form:

$$a_{mi} = \frac{W^0_{mi}}{V_m} \exp\left[-B \frac{T^2}{\beta^2} \log^2 \frac{p_0}{p}\right], \tag{5.9}$$

where:

$$B = \left(\frac{2.303 R}{E_0}\right)^2. \tag{5.10}$$

The publications of Dubinin [8–10] and many others demonstrate experimentally that these equations for adsorption of various vapours on active carbons of varying microporous structures are valid. The parameters W^0_{mi} and B of the D–R equation (5.9) are determined from the experimental adsorption isotherm in its linear form (Fig. 5.1) [11]:

$$\log a_{mi} = C - D \log^2 \frac{p_0}{p}, \tag{5.11}$$

where

$$C = \log \frac{W^0_{mi}}{V_m} \tag{5.12}$$

and

$$D = 0.434 \, B \frac{T^2}{\beta^2}. \tag{5.13}$$

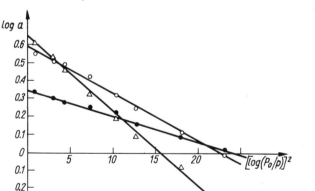

Fig. 5.1 Isotherms of benzene adsorption at 20°C in linear form according to Eq. (5.11) for Soviet active carbons: W-1 — full circles, W-2 — empty circles, W-3 — empty triangles (after Izotova and Dubinin [11]).

The D–R equation (5.9) is adhered to particularly well by the rather scarce microporous carbon adsorbents featuring a narrow distribution of the adsorption space. Strongly activated carbons have a wider distribution of the adsorption space of the micropores as a function of their dimensions which may be accurately described by the superposition of two distributions [11–13] i.e. of micropores (with effective radius $r < 0.6$ to 0.7 nm) and of supermicropores (0.6 to $0.7 < r < 1.5$ to 1.6 nm) [14]:

$$W = W_{mi}^{01} \exp\left[-\left(\frac{A}{\beta E_{01}}\right)^2\right] + W_{mi}^{02} \exp\left[-\left(\frac{A}{\beta E_{02}}\right)^2\right] \quad (5.14)$$

with:

$$W_{mi}^0 = W_{mi}^{01} + W_{mi}^{02}. \quad (5.15)$$

The parameters of equation (5.14), i.e. the limiting volume of the adsorption space and the characteristic adsorption energies E_{01} and E_{02} of the micropores W_{mi}^{01} and supermicropores W_{mi}^{02}, respectively, are easily determined graphically from one adsorption isotherm for a wide range of equilibrium pressures.

5.2 DEVELOPMENTS OF THE MICROPORE FILLING THEORY

Analysis of extensive experimental data concerning adsorption equilibria on active carbons with different parameters W_{mi}^0 and E_0 for the porous structure has shown that equation (5.5) conforms very well to the experimental data for characteristic energies of benzene adsorption $E_0 < 25$ kJ mol^{-1}. For active carbons with very fine micropores, when E_0 reaches a value of 30 kJ mol^{-1}, the upper limit of applicability of equation (5.5) is shifted towards small degrees of filling. For instance, Dubinin and Polstyanov [13] found that for Saran carbon, equation (5.5) is only formally applicable for benzene and cyclohexane in the range $\theta_{mi} = 0.1$ to 0.47 for effective values of W_{mi}^0 almost 1.5 times greater than the real volume of the micropores. Extension of the theory of volume filling of micropores to systems with stronger adsorption interactions, and especially systems with a high adsorption energy in the smallest micropores, has been based on application of the more general Weibull equation [15] for representing the distribution of the degree of filling θ_{mi} as a function of the differential molar work of adsorption A. In the case considered, the Weibull distribution function $F(A)$ in the normalized form is given by the equation:

$$F(A) = 1 - \exp\left[-\left(\frac{A}{E}\right)^n\right] \tag{5.16}$$

From this equation it follows after accounting for (5.2) that if $A = 0$, i.e. $p/p_0 = 1$, then $F(A) = 0$. However, if p/p_0 decreases in the range of very small equilibrium relative pressures, then A becomes very large, and $F(A)$ tends to unity. Hence $F(A)$ represents the unfilled volume of the micropore adsorption space $1 - \theta_{mi}$. We therefore have:

$$F(A) = 1 - \theta_{mi} = 1 - \left[\exp-\left(\frac{A}{E}\right)^n\right] \tag{5.17}$$

and

$$\theta_{mi} = \exp\left[-\left(\frac{A}{E}\right)^n\right]. \tag{5.18}$$

Equation (5.18) is known as the Dubinin–Astakhov (D–A) equation [16]. It contains, as compared with equation (5.5), a new parameter n characterizing the width of the micropore distribution function.

Kadlec [17] was first to assume that parameter n of equation (5.18) is the third variable parameter. This assumption drastically changed the properties of the equation and its application to microporous structures of carbon adsorbents. The works of Rand [18], our own experiments and the calculations in ref. [19] have confirmed the validity of characterizing microporous adsorbents by three parameters: W_{mi}^0 — the limiting volume of the adsorption space (volume of the micropores), E — the characteristic adsorption energy, and n — the exponent.

5.3 HETEROGENEITY OF THE MICROPOROUS STRUCTURE

The polydispersity of the microporous structure is usually regarded as a feature of its heterogeneity [20]. The wide distribution of the micropores has two major causes. The first is due to the heterogeneity of the starting material, i.e. bituminous coal or brown coal. Raw materials of vegetable origin chiefly consist of cellulose and lignin. The products of their carbonization reveal a differentiated chemical reactivity. Since coals contain various petrographic components such as fusain, vitrain, and clarain, they differ as regards their chemical vulnerability to the action of activating gases after thermal treatment (carbonization), and therefore each component is activated in a characteristic way.

The second cause lies in the heterogeneous nature of the activation process of the carbon particles or granules with gaseous substances. Even under laboratory conditions, when accurate mixing of the carbon is maintained during the activation process, access of the reacting gas to the carbon particles is never quite uniform. The gas penetrates into the particles via diffusion which is accompanied by chemical reaction. Hence, the concentration of the activating gas decreases with the distance from the particle surface. Since the activation of carbon is endothermic, the temperature in the reaction zone is lowered, apart from which heat penetrates very slowly to the deepest layers of the particle. All these factors make activation dependent on the carbon particle radius. If carbon activated by the gas-steam method is ground and size fractionated, then owing to the different mechanical strength of particles activated to different degrees, the most activated carbon will accumulate in the finest fractions. For these reasons every fraction has its specific adsorption isotherm [21]. These problems may be avoided by applying chemical (e.g. zinc chloride) activation of the carbon.

In terms of the VFM theory, the generation of heterogeneous microporous adsorbent structures originates in the creation of supermicropores during the activation process. The data presented by Dubinin [12] enable us to ascertain that the characteristic dimensions of micropores are 1.8–2.0 times greater than those of true micropores. Hence it can be concluded that the supermicropores are created stepwise (not gradually), most probably as a result of the merging of neighbouring micropores due to burn-off the separating walls.

5.4 FILLING OF HETEROGENEOUS MICROPOROUS STRUCTURES

Stoeckli [22] initiated the development of the theory of volume filling for heterogeneous microporous structures because the D–R equation (5.9) does not describe satisfactorily the process of adsorption on adsorbents with a wide micropore-size distribution. The D–R equation should be replaced for heterogeneous adsorbents by an expression appropriate to a structurally heterogeneous solid. The following integral equation, advanced by Stoeckli [22] in 1977, is used for describing the characteristic adsorption curves of single vapours of gases on heterogeneous microporous adsorbents:

Sec. 5.4] Filling of heterogeneous structures

$$\theta(A) = \int_\Delta \theta_{mi}(A, B) F(B) \, dB, \tag{5.19}$$

where $\theta(A)$ is the adsorption isotherm for a heterogeneous microporous adsorbent, $\theta_{mi}(A, B)$ is the local adsorption isotherm for homogeneous micropores characterized by the structural parameter B related to the characteristic adsorption energy E_0 for the standard vapour by equation (5.10), $F(B)$ is the normalized micropore distribution of the structural parameter B that characterizes the structural heterogeneity of the microporous adsorbent, and Δ is the range of integration.

Stoeckli [22] used as the $F(B)$ function the Gaussian distribution normalized over the range $\Delta = (-\infty, +\infty)$:

$$F(B) = \frac{1}{\sigma\sqrt{2\pi}} \exp\left[-\frac{(B_0 - B)^2}{2\sigma^2}\right] \tag{5.20}$$

where B_0 is the value of B at the distribution function maximum and σ is the dispersion. Note that the distribution in the range from $-\infty$ to 0 has no physical meaning and therefore the adsorption isotherms derived from the normalized Gaussian distribution (5.20) for the range $-\infty$ to $+\infty$ hold only for systems with fairly narrow micropore distributions. The physical condition requires that the function $F(B)$ equals zero when B is zero.

The integral equation (5.19), after substituting (5.9) for the local isotherm $\theta_{mi}(A, B)$ and (5.20) for the function $F(B)$ and after integration assumes the following form for the adsorption isotherm known as Dubinin–Radushkevich–Stoeckli (D–R–S) equation:

$$a_{mi} = a_{mi}^0 \exp(-B_0 y) \exp\left(y^2 \frac{\sigma^2}{2}\right) \frac{1 - \mathrm{erf}(z)}{2}, \tag{5.21}$$

where:

$$y = \left(\frac{T}{\beta} \log \frac{p_0}{p}\right)^2, \tag{5.22}$$

$$z = \left(y - \frac{B_0}{\sigma^2}\right) \frac{\sigma}{\sqrt{2}}, \tag{5.23}$$

and $\mathrm{erf}(z)$ is the tabulated error function. The D–R–S equation includes three parameters, namely a_{mi}^0 which is the limiting adsorption that corresponds to complete filling of the entire volume of the adsorption space of the heterogeneous micropores, B_0 which is the structural parameter at which the micropore distribution function $F(B)$ reaches its maximum, and σ which is the dispersion of that distribution. In refs. [22–25] it is claimed that equation (5.21) describes satisfactorily the experimental isotherms of vapours and gases of many substances at different temperatures over a range of relative pressure of 10^{-6} to 10^{-1}.

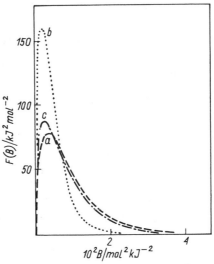

Fig. 5.2 Distribution of the $F(B)$ function calculated from equation (5.24) for the following active carbons: (a) AC Merck, (b) AG 5 Hajnówka (Poland), (c) M22 (laboratory carbon from apricot stones) (after Jaroniec and Choma [27]).

Some difficulties in the physical interpretation of the Gaussian distribution function for real microporous adsorbents and the relatively complex form of the D–R–S adsorption isotherm (5.21) prompted a search for other solutions. Jaroniec and Piotrowska [26] proposed the following gamma distribution function as the $F(B)$ function:

$$F(B) = \frac{q^{n+1}}{\Gamma(n+1)} \exp(-q^n), \qquad (5.24)$$

where: $q > 0$ and $n > -1$ are parameters of the Γ function. The $F(B)$ function (examples of which are given in Fig. 5.2) is characterized by the following values:

— maximum value

$$B_0 = \frac{n}{q}, \qquad (5.25)$$

— average value

$$\bar{B} = \frac{n+1}{q}, \qquad (5.26)$$

— dispersion

$$\sigma_B = \frac{\sqrt{n+1}}{q}. \qquad (5.27)$$

If we substitute the D–R equation for the local isotherm (5.9) and the micropore gamma distribution function (5.24) into the integral equation (5.19) we obtain the following equation for the adsorption isotherm:

$$\theta_{mi}(A) = \frac{a_{mi}}{a_{mi}^0} = \left[\frac{q}{q+(A/\beta)^2}\right]^{n+1}. \qquad (5.28)$$

This equation was derived for the first time in ref. [27] and is referred to as the Jaroniec–Choma (J–C) equation. It contains three parameters: a_{mi}^0 which is the limiting adsorption in the micropores, and q and n which are parameters characterizing the micropore distribution function $F(B)$. As shown in refs. [27–30], the J–C equation is very promising for the description of adsorption in micropores.

Equation (5.28) can be linearized:

$$\ln a_{mi} = \ln a_{mi}^0 - (n+1)\ln\left[1 + \frac{1}{q}\left(\frac{A}{\beta}\right)^2\right]. \qquad (5.29)$$

For large values of q (when the $F(B)$ distribution is relatively narrow and the adsorbent homogeneous) and sufficiently small values of A, equation (5.29) assumes the form:

$$\ln a_{mi} = \ln a_{mi}^0 - \bar{B}\left(\frac{A}{\beta}\right)^2 \qquad (5.30)$$

where $\bar{B}$ is the mean value of B (Eq. 5.26) for the gamma function. Equation (5.30) is the logarithmic form of the D–R equation (5.11); thus for homogeneous microporous adsorbents the J–C equation reduces to the D–R version [31].

Another possibility of solving the integral relation (5.19) is to apply the D–A equation (5.18) as the isotherm describing adsorption in micropores of specified dimensions (local isotherm) $\theta_{mi}(A, B)$. This approach was applied by Rozwadowski et al. [32, 33] who obtained, on the basis of equations (5.18), (5.19) and (5.20), the following equation for the adsorption isotherm:

$$\theta_{mi}(A) = \frac{a_{mi}}{a_{mi}^0} = \exp\left[\frac{\sigma^2}{2}\left(\frac{A}{\beta}\right)^{2n'}\right] \exp\left[-k_0\left(\frac{A}{\beta}\right)^{n'}\right] \times \frac{\operatorname{erfc}\left[\frac{\sigma}{\sqrt{2}}\left(\frac{A}{\beta}\right)^{n'} - \frac{k_0}{\sigma\sqrt{2}}\right]}{\operatorname{erfc}\left(-\frac{k_0}{\sigma\sqrt{2}}\right)}, \quad (5.31)$$

where k_0 is the maximum of the micropore volume distribution function, n' is a constant equal to 2 or 3 depending on the dimensions of the micropores, and $\operatorname{erfc}(x)$ is the complement error function.

If we assume that the micropores in adsorbents with heterogeneous microporous structures are flat slits of limited length, then we can calculate the basic parameter of the assumed model, i.e. half of the slit width x. This quantity is related to the characteristic energy of adsorption of benzene vapour as follows:

$$x = \frac{k}{E_0}, \quad (5.32)$$

where x is given in nm, E_0 in kJ mol^{-1}, and k in kJ nm mol^{-1}. The parameter k depends only slightly on E_0 so it can be found [25] from the relation:

$$k = 13.028 - 1.53 \times 10^{-5} E_0^{3.5}. \quad (5.33)$$

The characteristic energy of adsorption E_0 in kJ mol^{-1} is related to the structural parameter B in mol^2 kJ^{-2} by the equation:

$$E_0 = \sqrt{\frac{1}{B}}. \quad (5.34)$$

From equations (5.32) and (5.34) we obtain:

$$B = cx^2 \quad (5.35)$$

where:

$$c = \left(\frac{1}{k}\right)^2. \quad (5.36)$$

For relatively large micropores c is constant and equals 0.006944 $(kJ^{-1} nm^{-1} mol)^2$. Expression (5.35) was derived for the first time by Dubinin [34] on the basis of an exacting interpretation of experimental data. From this equation we see that the distribution function $F(B)$ can be transformed into the micropore volume distribution function $G(x)$ where x is a linear dimension [35]:

$$G(x) = 2cx F[B(x)]. \tag{5.37}$$

The final form of the micropore volume distribution as a function of their dimensions, $G(x)$, associated with the distribution function $F(B)$ (5.24) is:

$$G(x) = \frac{2(q_c)^{n+1}}{\Gamma(n+1)} x^{2n+1} \exp(-q_c x^2), \tag{5.38}$$

where

$$q_c = qc, \tag{5.39}$$

q and n being the parameters of the J–C adsorption isotherm (5.28).

Another equation that seems promising for the description of adsorption in micropores is derived by applying the $F(B)$ distribution function (5.20) based on the Gaussian micropore distribution function [36]:

$$G(x) = \frac{1}{\sigma'^2 \sqrt{2\pi}} \exp\left[-\frac{(x-x_0)^2}{2\sigma'^2}\right], \tag{5.40}$$

where x_0 corresponds to the maximum of the distribution function, and σ' is the dispersion. This equation, known as the Dubinin–Stoeckli (D–S) adsorption isotherm, has the following form:

$$a_{mi} = \frac{a_{mi}^0}{2\sqrt{Y(A)}} \exp\left[-\frac{mx_0^2 A^2}{Y(A)}\right]\left[1+\mathrm{erf}\left(\frac{x_0}{\sigma'\sqrt{2Y(A)}}\right)\right], \tag{5.41}$$

where:

$$Y(A) = 1 + 2m\sigma'^2 A^2, \tag{5.42}$$

$$m = \frac{c}{\beta^2}, \tag{5.43}$$

and erf is the error function. The parameters of the D–S equation are a_{mi}^0 which is the maximum adsorption in the micropores, x_0 which is the value of x at the maximum of the distribution function described by (5.40), and σ' which is the dispersion of that distribution.

Table 5.1

Parameters of the microporous structure of active carbons calculated from the J–C equation (5.28) and the D–S equation (5.41) using the benzene adsorption isotherm at 20°C (after Jaroniec and Choma [29])

Active carbon	Parameters of the J–C equation (5.28)				Parameter of the D–S equation (5.41)			
	a_{mi}^0 /mmol g^{-1}	n	q /kJ^2 mol^{-2}	SD	a_{mi}^0 /mmol g^{-1}	x_0 /nm	σ' /nm	SD
AC	5.21	0.86	481	0.178	5.20	0.69	0.26	0.171
BH	2.67	0.21	97	0.104	2.68	1.15	0.55	0.116

Data in Table 5.1 provide examples of the characterization of the microporous structure of two active carbons by means of the J–C isotherm (5.28) and of the D–S isotherm (5.41). The standard deviation (SD) included in the table show that these equations describe equally well the experimental data.

The function $G(x)$ for the micropore-size distribution corresponds to the J–C adsorption isotherm equation (5.28), while the D–S equation follows directly from the distribution function $G(x)$ described by equation (5.40). Despite the significantly different analytical forms of these function, they generate for real microporous active carbons very

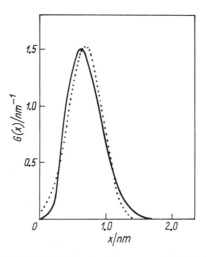

Fig. 5.3 The $G(x)$ distribution of micropores calculated from equation (5.38) (solid line) and from equation (5.40) (dotted line) for active carbon AC, the microporous structure parameters of which are given in Table 5.1 (after Jaroniec and Choma [29]).

similar curves. This is seen from Fig. 5.3. The distribution functions $G(x)$ give precise information about the structural heterogeneity of microporous adsorbents.

The theory of volume filling of micropores is now generally accepted and used for describing adsorption phenomena occurring in the micropores of adsorbents. The determination of the parameters of the D–R (5.9), D–A (5.18), D–R–S (5.21), J–C (5.28) and D–S (5.41) equations for many adsorbents simultaneously would be a burdensome task if we did not have available the algorithms for the numerical analysis of adsorption isotherms from the gas phase published in refs. [19, 37–40] which enable the determination of the parameters of the porous structure of adsorbents on the basis of the TVFM.

5.5 SURFACE AREA OF THE MICROPORES

The TVFM provides also a method for calculating the surface area of micropores on the basis of adsorption data. If we know the parameters of equations (5.9) and (5.18) for carbon adsorbents and the limiting volume of the adsorption space, i.e. the volume W_{mi}^0 of the micropores and the characteristic energy of adsorption E_0, we can easily calculate the geometric surface area of the micropore walls for the slit-like model [41, 42]:

$$S_G = \frac{W_{mi}^0}{x_0} \times 10^3 \; m^2 \; g^{-1}. \tag{5.44}$$

The volume W_{mi}^0 of the micropores is the total volume of the flat slits of half-width x_0 (W_{mi}^0 is in cm^3 g^{-1} and x_0 in nm).

The geometric surface area of active carbon pores for a wide micropore distribution, satisfactorily described by (5.14), can be calculated [25] from equation:

$$S_G = \left(\frac{W_{mi}^{01}}{x_{01}} + \frac{W_{mi}^{02}}{x_{02}}\right) \times 10^3 \; m^2 \; g^{-1}, \tag{5.45}$$

where W_{mi}^{01}, x_{01} and W_{mi}^{02}, x_{02} are the characteristic parameters of the true micropores and supermicropores, respectively.

The D–R–S equation also enables determination of the micropore surface area. Every increment dW_0 is associated with an increase of the geometric surface area of the micropore walls by dS_G:

$$dS_G = \frac{dW_{mi}^0}{x} \times 10^3 \tag{5.46}$$

According to (5.40) and (5.46) we have:

$$dS_G = \frac{2000\, c W^0_{mi}}{\sigma'\sqrt{2\pi}} \exp\left[-\frac{c^2(x_0^2-x^2)^2}{2\sigma'^2}\right], \tag{5.47}$$

where W^0_{mi} is the total pore volume.
Putting

$$D = \frac{2000\, c W^0_{mi}}{\sigma'\sqrt{2\pi}}, \tag{5.48}$$

$$F = \frac{c^2}{2\sigma'^2}, \tag{5.49}$$

$$H = x_0^2, \tag{5.50}$$

and integrating, we obtain:

$$S_G = D \int_a^b \exp[-F(H-x^2)^2]\, dx. \tag{5.51}$$

Another method, independent of the TVFM method of determining the geometric surface area of microporous carbon adsorbents, is that based on the adsorption of water vapour due to the formation of hydrogen bonds. In this case the dispersive interaction is small and to a first approximation may be neglected. Then the adsorption on the walls of slit-like micropores proceeds in the same way as on the surface of a nonporous carbon adsorbent. Thus both for carbon black-950 with a specific surface area of 91 m² g⁻¹ [25] and for microporous adsorbents of chemically similar surface character, the monolayer of adsorbed water molecules is formed at 20°C for an equilibrium relative pressure of $p/p_0 = 0.6$ [41]. Since for carbon black-950 the amount of monomolecular adsorbed (a_m) under these conditions reaches 1.44 mmol g⁻¹, we can write [42]:

$$\frac{a_{0.6}}{1.44} = \frac{S_G^{H_2O}}{91} \tag{5.52}$$

and

$$S_G^{H_2O} = \frac{91}{1.44} a_{0.6} = 63.2\, a_{0.6}, \tag{5.53}$$

where $a_{0.6}$ is the adsorption of water for a microporous carbon adsorbent at $p/p_0 = 0.6$ and $S_G^{H_2O}$ is the geometrical surface area of the micropore walls found from the water vapour adsorption isotherm.

Sec. 5.6]   **Thermodynamics of adsorption in micropores**   155

In the theory of volume filling of micropores, one retains a physical picture of the geometric surface area of the walls of the carbonaceous adsorbent micropores as determined by one of the methods given above. The application of methods based on the assumption of multilayer adsorption such as the BET [43] or de Boer t-methods [44] for determination of the specific area of microporous adsorbents leads, in contrast to nonporous or macroporous materials, to the derivation of values much greater than those found on the basis of the TVFM. This is seen from the data given in Table 5.2 where the values of x_0 and S_G have

Table 5.2
Specific surface areas of model microporous carbon adsorbents for benzene vapours at 20°C, $W_{mi}^0 = 0.5$ cm^3 g^{-1} (after Dubinin [41])

E_0 /kJ mol^{-1}	x_0 /nm	S_G	S_{BET}	S_t
			/m^2 g^{-1}	
18.9	0.67	750	1200	1450
13.3	0.97	520	1170	1280
9.4	1.38	360	1040	1050

been determined for the parameters W_{mi}^0 and E_0 were calculated according to the D–R equation (5.9) in terms of the BET and de Boer methods. It follows from the TVFM that the use of adsorption methods based on the assumption of multilayer adsorption, in terms of which successive adsorption layers are formed, in the determination of the specific surface area of microporous adsorbents is unjustified, since the specific surface areas found in this way have no physical meaning [42]. However, this conclusion, following from the TVFM, is not universally accepted as we have noted in Chapter 4.

5.6 THERMODYNAMICS OF ADSORPTION IN MICROPORES

Bering and Serpinskii [45–48] have considered some thermodynamic aspects of the TVFM. They defined thermodynamic criteria of the fundamental assumption of the theory, i.e. the temperature-independence of the characteristic adsorption curve and that of the differential molar heat and entropy of adsorption.

Let adsorption at temperature T and pressure p be a. If in the calculations of the thermodynamic functions of the adsorbate, we assume the liquid phase at the same temperature as an arbitrarily chosen standard system, then the maximum work A of transporting one mole of adsorbate from the liquid phase to an infinitely large amount of adsorbent is given by equation (5.2). The differential molar entropy of the adsorbate ΔS, calculated with respect to the same reference state, is given by the expression:

$$\Delta S = -\left(\frac{\partial \Delta G}{\partial T}\right)_a = \left(\frac{\partial A}{\partial T}\right)_a \tag{5.54}$$

and the change of enthalpy ΔH, equal to the differential heat q of adsorption, is:

$$\Delta H = -q = \Delta G + T\Delta S = -A + T\left(\frac{\partial A}{\partial T}\right)_a. \tag{5.55}$$

The general expression for the differential molar heat of adsorption has the form [49]:

$$Q = \lambda + A - T\left(\frac{\partial A}{\partial T}\right)_\theta - \alpha T\left(\frac{\partial A}{\partial \ln a}\right)_T, \tag{5.56}$$

where: α — the limiting adsorption temperature coefficient — is:

$$\alpha = -\frac{\partial \ln a_{mi}^0}{\partial T} \tag{5.57}$$

and λ is the heat of condensation of the adsorbate. The 'net' differential heat of adsorption q is given by the equation:

$$q = Q - \lambda. \tag{5.58}$$

In the case of adsorption in micropores, when the condition of temperature-independence is fulfilled:

$$\left(\frac{\partial A}{\partial T}\right)_\theta = 0 \tag{5.59}$$

and equations (5.54) and (5.56) take the form:

$$\Delta S = -\alpha RT\left(\frac{\partial \ln h}{\partial \ln a}\right)_T, \tag{5.60}$$

$$Q = \lambda - RT \ln h + \alpha RT^2\left(\frac{\partial \ln h}{\partial \ln a}\right)_T, \tag{5.61}$$

Sec. 5.6]   Thermodynamics of adsorption in micropores

where

$$h = \frac{p}{p_0}.$$

Consider the two principal cases in the calculation of differential heats and entropies of adsorption. The first and commonest case involves the use of the one-term D–A adsorption equation (5.18) for microporous adsorbents for which the effect of dispersion forces is greatest. The 'net' differential heat of adsorption for the standard vapour is given by the equation [50, 51]

$$q = A + \frac{\alpha T E_0}{n} \left(\ln \frac{a^0_{mi}}{a} \right)^{\frac{1}{n}-1}, \tag{5.62}$$

where a^0_{mi} is the limiting value of adsorption, or in the form of the explicit function A [52]:

$$q = A + \frac{\alpha T}{n A^{n-1}} E_0^n. \tag{5.63}$$

The equation for the differential entropy has in this case the form:

$$\Delta S = -\frac{\alpha E_0}{n} \left(\ln \frac{a^0_{mi}}{a} \right)^{\frac{1}{n}-1}. \tag{5.64}$$

The differential molar heat of adsorption of benzene as a function of adsorption on active carbons is given in Fig. 5.4.

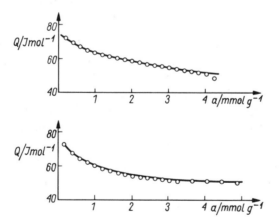

Fig. 5.4 Variation of the differential heat of adsorption of benzene with the magnitude of adsorption for two different active carbons. Solid lines — calculated heats, circles — experimental points (after Dubinin and Polstyanov [53]).

The second case relates to adsorbents with micropores and supermicropores. Here the equation of the 'net' differential heat of adsorption takes the form:

$$q = A + \frac{\alpha T}{n A^{n-1}} \left[\gamma E_{01}^n + (1-\gamma) E_{02}^n \right] \qquad (5.65)$$

where

$$\gamma = \frac{W_{mi}^{01}}{W_{mi}^{01} + W_{mi}^{02}}. \qquad (5.66)$$

The calculation of q from the experimentally determined differential molar heat of adsorption Q requires knowledge of the heat of condensation λ which amounts for benzene to 33.94 kJ mol^{-1} [52].

Concluding we should note that the TVFM not only enables us to describe the adsorption equilibrium over a fairly wide range of temperatures and pressures but also to calculate with sufficient accuracy, especially for practical purposes, the differential heat and entropy of adsorption. Of course such calculations are reliable only for the extent of coverage of the adsorbent for which the initial assumptions of the theory hold. It is significant that the parameters of the equations can be derived from a small amount of experimental data.

5.7 COMPLETE ANALYSIS OF THE POROUS STRUCTURE OF ADSORBENTS

All the approaches considered so far to adsorption phenomena based on the TVFM enable satisfactory determination of the parameters of the porous structure of the microporous adsorbents tested. These approaches relate, however, only to microporous structures and do not extend to adsorption processes taking place in meso- and macropores. An important advance in this field was made by comparing adsorption and porosimetric data. Attempts to solve this problem due to Kadlec [54–57], depend on completely new pore distribution function of the following form:

$$V(r) = V_0 \exp\left[-R^{-h} \exp \varphi(r)\right] \qquad (5.67)$$

where $V(r)$ is the total volume of pores of radii $\leqslant r$, R equals r/r_m, and the function $\varphi(r)$ is given for active carbons by the equation:

$$\varphi(r) = k_c |R'[1-(1+c_2|R'|)^{-1}]| \qquad (5.68)$$

where

$$k_c = c_1 c_2^{-1}\left(1+\frac{c_2}{h}\right) \qquad (5.69)$$

and

$$R' = 1 - R \qquad (5.70)$$

The parameters of equation (5.67) have a precise physical meaning, thus V_0 is the total volume of pores (micro, meso or macro), r_m is the characteristic radius of the pores, h is a parameter describing the width of the differential distribution function $V'(r)$, and c_1 and c_2 are parameters related to the asymmetry specific for the distribution function. The variation of the pore volume as a function of the logarithm of the pore radius for active carbon is shown in Fig. 5.5.

The determination of the parameters V_0, r_m, h, c_1 and c_2 has been described in detail in ref. [58], where an algorithm for numerical determination of the parameters of micro-, meso-, and macropores of the adsorbents is proposed. The parameters characterizing the carbon adsorbents considered in ref. [58], which are also those of the Kadlec equation (5.67), are summarized in Tab. 5.3.

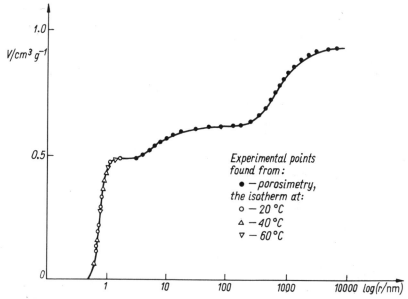

Fig. 5.5 The integral distribution of the pore volume V as a function of the common logarithm of the pore radius r for Norit RKD-3 active carbon (after Kadlec [55]).

Table 5.3

Computer-derived parameters of the porous structure of active carbons as determined by the Kadlec method (after Kadlec et al. [58])

Active carbon	Type of pores	V_0 /cm^3 g^{-1}	r_m /nm	h	c_1	c_2	S /m^2 g^{-1}
HS-43	micro	0.6357	0.769	3.875	0.910	2.450	1544
	meso	0.1899	74.3	1.230	−0.180	20.45	5.09
	macro	0.2247	4157.9	1.130	0.000	20.00	0.11
M-22	micro	0.6417	0.770	4.516	−4.910	0.450	1606
	meso	0.5046	4.786	2.136	0.450	19.55	181.1
	macro	0.5857	653.2	0.509	0.000	19.45	2.61
AG-50	micro	0.4574	0.717	5.232	2.890	2.450	1108
	meso	0.1867	7.740	0.887	−0.110	19.45	30.89
	macro	0.5524	1167.9	1.008	−0.060	20.45	1.02

From the parameters of the integral pore distribution function we can determine the differential distribution with the help of the equation:

$$\frac{dV}{d\ln r} = V(r)\left\{R^{-h}\exp\varphi(r)\times\right.$$

$$\left.\times\left[h - k_c R\left(\frac{1+c_2|R'|-c_2 R' \text{ sign } R'}{(1+c_2|R'|)^2} - 1\right)\text{sign } R'\right]\right\}. \quad (5.71)$$

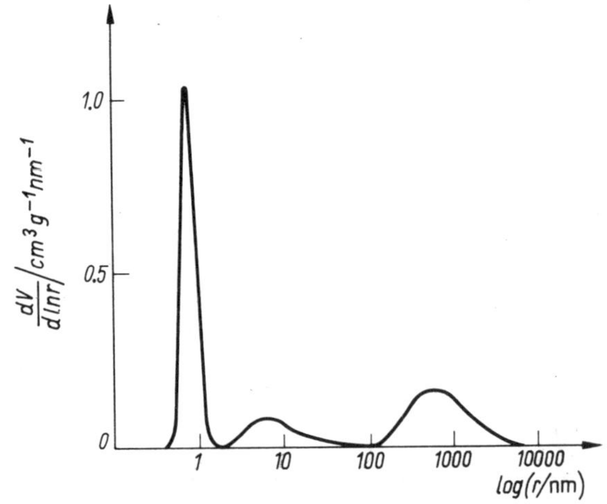

Fig. 5.6 The differential distribution of the pore volume as a function of the common logarithm of the effective pore radius r as calculated from equation (5.71) for Norit RKD-3 active carbon (after Kadlec [55]).

The differential distribution of the pore volume as a function of the logarithm of the radius for Norit active carbon is presented in Fig. 5.6. The method of characterizing the texture of the adsorbent would be incomplete if the specific surface area of the particular types of pores was unknown. According to Kadlec, these surface areas are determined from the equations:

$$S_{micro} = B \int_0^{V_{micro}} \frac{1}{r} dV \qquad (5.72)$$

$$S_{meso} = B \int_0^{V_{meso}} \frac{1}{r} dV \qquad (5.73)$$

$$S_{macro} = B \int_0^{V_{macro}} \frac{1}{r} dV \qquad (5.74)$$

where B is a constant related to the shape of the pores (e.g. for bottomless cylindrical pores $B = 2$).

The method of complete analysis of the porous structure does not provide an absolute and definitive account of the porous structure of adsorbents. The word 'complete' means that all three types of pores, i.e. micro-, meso- and macro-, are considered.

As seen from this survey, none of the methods gives the final solution to the problem posed.

References

[1] Dubinin, M. M., *Usp. Khim*, **21**, 513 (1952).
[2] Dubinin, M. M., *ibid.* **24**, 3 (1955).
[3] Dubinin, M. M., Zaverina, E. D., Radushkevich, L. V., *Zh. Fiz. Khim.* **21**, 1351 (1947).
[4] Dubinin, M. M., Zaverina, E. D., Timofeev, D. P., *ibid.* **23**, 1129 (1949).
[5] Dubinin, M. M., Zaverina, E. D., *ibid.* **24**, 1410 (1950).
[6] Nikolaev, K. M., Dubinin, M. M., *Izv. Akad. Nauk SSSR, O. Kh. N.*, 1165 (1958).
[7] Dubinin, M. M., Zaverina, E. D., Timofeev, D. P., *ibid.* 670 (1967).
[8] Dubinin, M. M., *Zh. Fiz. Khim.* **39**, 1305 (1965).
[9] Dubinin, M. M., *Chemistry and Physics of Carbon*, **2**, 51, M. Dekker, New York 1966.
[10] Dubinin, M. M., Carbon **13**, 263 **(1975)**.
[11] Izotova, T. I., Dubinin, M. M., *Zh. Fiz. Khim.* **39**, 2796 (1965).
[12] Dubinin, M. M., *Izv. Akad. Nauk SSSR, Ser. Khim.* 1691 (1979).
[13] Dubinin, M. M., Polstyanov, E. F., *ibid.* 793 (1966).
[14] Dubinin, M. M., *ibid.* 996 (1974).
[15] Weibull, W., *J. Appl. Mech.* **18**, 293 (1951).
[16] Dubinin, M. M., Astakhov, V. A., *Izv. Akad. Nauk SSSR, Ser. Khim.* 5 (1971).
[17] Kadlec, O., *Chem. Zvesti* **29**, 660 (1975).
[18] Rand, B., *J. Colloid. Interface Sci.* **56**, 337 (1976).

[19] Jankowska, H., Choma, J., *Przem. Chem.* **60**, 450 (1981).
[20] Dubinin, M. M., *Izv. Akad. Nauk. SSSR, Ser. Khim.* **18** (1980).
[21] Jankowska, H., Świątkowski, A., Witkiewicz, Z. et. al., *Biul. WAT*, **26**, No. 1, 42 (1977).
[22] Stoeckli, H. F., *J. Colloid Interface Sci.* **59**, 184 (1977).
[23] Huber, U., Stoeckli, H. F., Houriet, J. Ph., *ibid.* **67**, 195 (1978).
[24] Dubinin, M. M., *Carbon* **19**, 321 (1981).
[25] Dubinin, M. M., Efremov, S. N., Kataeva, L. I. et. al., *Izv. Akad. Nauk. SSSR, Ser. Khim.* 1711 (1981).
[26] Jaroniec, M., Piotrowska, J., *Monatsh. Chem.* **117**, 7 (1986).
[27] Jaroniec, M., Choma, J., *Materials Chem. Phys.* **15**, 521 (1986).
[28] Choma, J., Jankowska, H., Piotrowska, H. et. al., *Monatsh. Chem.* **118**, 315 (1987).
[29] Jaroniec, M., Choma, J., *Materials Chem. Phys.* **18**, 103 (1987).
[30] Choma, J., Jaroniec, M., Piotrowska, J., *Carbon* **26**, 1 (1988).
[31] Choma, J., Jaroniec, M., *Chem. Stosowana*, **38**, 385 (1988).
[32] Rozwadowski, M. Wojsz, R., *Carbon* **22**, 363 (1984).
[33] Rozwadowski, M., Wojsz, R., *ibid.* **22**, 437 (1984).
[34] Dubinin, M. M., *Characterization of Porous Solids* (Ed. Gregg, S. J., Sing, K. S. W., Stoeckli, H. F.), Society of Chemical Industry, London, 1–11, 1979.
[35] Dubinin, M. M., Stoeckli, H. F., *J. Colloid Interface Sci.* **75**, 34 (1980).
[36] Dubinin, M. M., *Carbon* **23**, 373 (1985).
[37] Lezin, U. S., Dubinin, M. M., *Dokl. Akad. Nauk SSSR* 1104 (1966).
[38] Dubinin, M. M., Lezin, U. S., Kadlec, O. et al., *Zh. Fiz. Khim.* **42**, 964 (1968).
[39] Brzeski, P., Choma, J., Pietrzyk, S., *Biul. WAT* **30**, No. 4, 103 (1981).
[40] Choma, J., *Przem. Chem.* **62**, 634 (1983).
[41] Dubinin, M. M., *Carbon* **18**, 355 (1980).
[42] Dubinin, M. M., *Izv. Akad. Nauk. SSSR, Ser. Khim.* 9 (1981).
[43] Brunauer, S., Emmett, P., Teller, E., *J. Am. Chem. Soc.* **60**, 309 (1938).
[44] De Boer, J. H., Broekhoff, J. C., Linsen, B. G. et al., *J. Catalysis*, **7**, 135 (1967).
[45] Bering, B. P., Serpinskii, V. V., *Dokl. Akad. Nauk SSSR* **114**, 1254 (1957).
[46] Bering, B. P., Serpinskii, V. V., *ibid.* **148**, 1331 (1963).
[47] Bering, B. P., Serpinskii, V. V., *Izv. Akad. Nauk. SSSR, Ser. Khim.* 847 (1971).
[48] Bering, B. P., Dubinin, M. M., Serpinskii, V. V., *J. Colloid Interface Sci.* **21**, 378 (1966).
[49] Dubinin, M. M., *Adsorption and Porosity*, Military Technical Academy, Warsaw 1975.
[50] Dubinin, M. M., *Progr. Surface Membrane Sci.* **9**, 1 (1975).
[51] Bering, B. P., Gordev, V. A., Dubinin, M. M. et al., *Izv. Akad. Nauk SSSR, Ser. Khim.* 22 (1971).
[52] Dubinin, M. M., Isirikyan, A. A., *ibid.* 13 (1980).
[53] Dubinin, M. M., Polstyanov, J. P., *ibid.* 1507 (1966).
[54] Kadlec, O., *Porozimetrie a jeji pouziti* **3**, 7 (1976).
[55] Kadlec, O., *ibid.* **4**, 7 (1977).
[56] Kadlec, O., *The Characterization of Porous Solids*, Proceedings Swiss-British Symposium, Neuchâtel 1978, London SCI 1979.
[57] Kadlec, O., *Porozimetrie a jeji pouziti* **5**, 9 (1979).
[58] Kadlec, O., Choma, J., Świątkowski, A., *Collect. Czechoslov. Chem. Comm.* **49**, 2721 (1984).

CHAPTER 6

Energy Effects in Adsorption

6.1. HEAT OF ADSORPTION

That adsorption is a spontaneous exothermic process is easily proved experimentally and confirmed thermodynamically.

If in a system a constant temperature and pressure are maintained, a spontaneous process must involve a decrease in the thermodynamic potential, i.e. $\Delta G < 0$. The transfer of adsorbate molecules from the gas or liquid phase to the adsorbent surface is accompanied by their ordering, i.e. by lowering their entropy, $\Delta S < 0$. In the case of an isothermal process,

$$\Delta G = \Delta H - T\Delta S. \qquad (6.1)$$

Since $\Delta G < 0$ (spontaneous process) and $T\Delta S < 0$, the change in enthalpy is always $\Delta H < 0$. This demonstrates that the process of adsorption is always exothermic.

The heat effect of the process is one of the criteria enabling a distinction to be made between physical and chemical adsorption on a solid adsorbent. In the case of physical adsorption the amount of energy released lies in the range 4.0 to 80 kJ mol^{-1}, while in the case of chemisorption the energy release is comparable to the thermal effects of chemical reactions and amounts to 40–400 kJ mol^{-1}. The absolute value of the heat of adsorption per mole of adsorbate decreases with increased coverage of the adsorbent surface.

The energy effects accompanying adsorption are considered below.

The *molar heat of adsorption* at constant pressure and temperature is given by the equation:

$$Q_i = H_{i(surf)} - H_{i(g)}, \qquad (6.2)$$

where $H_{i(surf)}$ is the mean molar enthalpy of the ith adsorbate on the adsorbent surface and $H_{i(g)}$ is the molar enthalpy of the same adsorbate in the gas phase.

Let $G_{i(surf)}$ be the chemical potential (partial molar thermodynamic potential) of the ith adsorbate in the surface layer, $G_{i(g)}$ be the same potential of the ith substance in the gas phase, and a_i the amount adsorbed of the ith substance.

At equilibrium:

$$G_{i(surf)} = G_{i(g)}, \tag{6.3}$$

$$dG_{i(surf)} = dG_{i(g)}.$$

Since $G_{i(surf)} = f(T, a_i)$ and $G_{i(g)} = \varphi(T, p)$, we have:

$$\frac{\partial G_{i(surf)}}{\partial T} dT + \frac{\partial G_{i(surf)}}{\partial a_i} da_i = \frac{\partial G_{i(g)}}{\partial T} dT + \frac{\partial G_{i(g)}}{\partial p} dp. \tag{6.4}$$

If we permit changes of temperature but assume the adsorbed amount to be constant (a_i = const) we obtain:

$$-S_{i(surf)} dT = -S_{i(g)} dT + V_{i(g)} dp, \tag{6.5}$$

where $S_{i(surf)}$ is the partial molar entropy of the ith substance in the surface layer. Equation (6.5) holds since

$$\frac{\partial G_{i(surf)}}{\partial T} = -S_{i(surf)}, \qquad \frac{\partial G_{i(g)}}{\partial T} = -S_{i(g)}, \tag{6.6}$$

$$\frac{\partial G_{i(surf)}}{\partial a_i} = 0 \quad \text{and} \quad \frac{\partial G_{i(g)}}{\partial p} = V_{i(g)}, \tag{6.7}$$

where $V_{i(g)}$ is the molar volume of the gaseous adsorbate. Transforming this equation we obtain:

$$S_{i(surf)} - S_{i(g)} = -V_{i(g)} \left(\frac{\partial p}{\partial T}\right)_{a_i}. \tag{6.8}$$

Since we are considering a system at equilibrium for which the condition T = const is always fulfilled, the following equation holds

$$S_{i(surf)} - S_{i(g)} = \frac{Q_i}{T}. \tag{6.9}$$

If we assume that the adsorbate in the gas phase shows ideal behaviour, then:

$$V_{i(g)} = \frac{RT}{p}. \tag{6.10}$$

Substituting equations (6.9) and (6.10) into (6.8) we obtain:

$$-Q_i = RT^2 \left(\frac{\partial \ln p}{\partial T}\right)_{a_i}. \qquad (6.11)$$

This equation has a form analogous to the Clausius–Clapeyron equation, which describes the change in the state of aggregation of chemical species under equilibrium conditions. The quantity Q_i in equation (6.11) is the *differential heat of adsorption*. The variation of this quantity with the degree of adsorption is shown in Fig. 6.1.

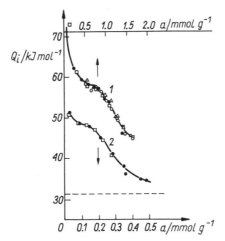

Fig. 6.1 Variation of the differential heat of adsorption of *n*-hexane at 20°C with degree of adsorption of this compound: 1 — on active carbon, 2 — on non-porous carbon black (heated preliminarily under vacuum at 950°C) (after Avgul' *et al.* [1]).

The isosteric heat of adsorption is found also from equation (6.11). For this purpose adsorption isotherms are measured for the given adsorbate–adsorbent system at several temperatures. Then points corresponding to a particular surface coverage (i.e. a definite amount adsorbed per unit mass of the adsorbent) are plotted in the $\ln p$ vs. $\frac{1}{T}$-coordinates. These points lie along straight lines which are known as the adsorption isosteres. The slopes of the isosteres are heats Q_i, known as *isosteric heats of adsorption*.

The isosteric heat of adsorption can also be found by a gas chromatography method. In this case the temperature dependence of the retention times or volumes is utilized. Full details can be found in the monograph by Paryjczak [2], where many examples of testing gas–solid adsorption systems are given (including active carbon systems).

The total heat of adsorption Q (referred to 1 mole of adsorbate in the surface layer) in the isothermal process associated with the formation of an adsorption layer corresponding to an amount adsorbed a_i and 1 g of adsorbent is found from the equation:

$$Q = \frac{1}{a_i} \int_0^{a_i} Q_i \, dn_i, \qquad (6.12)$$

where dn_i is the differential of the number of moles of adsorbate transferred from the gas phase to the adsorption layer.

By transforming equation (6.12) we obtain:

$$Qa_i = \int_0^{a_i} Q_i \, dn_i. \qquad (6.13)$$

Since

$$da_i = \frac{dn_i}{m}, \qquad (6.14)$$

where m is the mass of the adsorbent, we obtain:

$$Q_i = \frac{\partial}{\partial n_i}(a_i Q) = \left(Q + a_i \frac{dQ}{da_i}\right)\frac{1}{m} \qquad (6.15)$$

putting $m = 1$ g we finally obtain:

$$Q_i = Q + a_i \frac{dQ}{da_i}, \qquad (6.16)$$

where Q is the total heat of adsorption. This can be found from a calorimetric measurement, when the heat effect during the contact of a given quantity of adsorbate with the clean surface of the adsorbent is determined directly. Sometimes it is more convenient to determine the heat of wetting of the adsorbent with a liquid adsorbate. In the wetting process of a clean solid surface by a liquid, the total heat of adsorption may be determined. Next, the adsorbent is kept until equilibrium is achieved in the vapour of the adsorbate when the heat accompanying its contact with the liquid adsorbent is determined once more. In the latter case the heat effect will be equal to the difference between the total heat of adsorption and the heat of wetting. Hence we find the total heat of adsorption of the liquid.

Measurements of the heat of wetting have recently found application in the direct determination of the micropore distribution for active carbons. Stoeckli and Kraehenbuehl [3, 4] have shown that the parameters E_0 and W_0 of the D–R equation are related to the enthalpy of immersion in the micropores by the expression:

$$-\Delta H_i = \frac{\beta E_0 W_0 \sqrt{\pi}}{2 V_m}(1+\alpha T), \qquad (6.17)$$

where α is the thermal coefficient of the wetting liquid. For a wetting liquid with critical dimension of its molecules L, after account is also taken of the external surface area S_e, equation (6.17) can be written in the form:

$$W(L) = \frac{-(\Delta H_i - h_i S_e) 2 V_m}{\beta E_0 \sqrt{\pi}\,(1+\alpha T)}, \qquad (6.18)$$

where h_i is the specific enthalpy of wetting of the surface S_e. The possibilities of using immersion calorimetry for characterizing active carbons have been presented by Stoeckli [5].

6.2 SOME TYPES OF CALORIMETERS

Since knowledge of the thermal effects accompanying adsorption of gases and vapours on the surface of porous solids is very important for an improved understanding of molecular interactions in the adsorbate–adsorbent system, there has been a systematic development in calorimetric techniques designed for studying adsorption processes. In the last few decades (especially since 1940) many new designs of calorimeters for measuring the heats of adsorption have been developed that differ both as regards their principle of operation and their construction. Since many monographs [6–9] detail descriptions of the design and operation of various calorimeters, we shall confine ourselves to a general discussion of the principal types, the criteria for their classification, and design. Depending on what proportion of the total thermal effect is exchanged between the vessel containing the adsorbent and the environment, we can classify the calorimeters into the following main groups, adiabatic, isothermal, and diathermal.

The first group — adiabatic calorimeters — are designed to prevent the transfer of the heat released in the adsorption process to the environment. Prevention of energy exchange in the form of heat produces, in effect, an increase of temperature in the vessel with the adsorbent. The direct, accurate determination of this increase of temperature makes it possible to estimate the thermal effect accompanying adsorption. In order to prevent heat loss such calorimeters are provided with vacuum jackets and compensating electrical heaters. The measurement of temperature is carried out with thermocouples.

The calorimeters classified as the second group, i.e. isothermal calorimeters, are designed so that the heat released is completely transferred from the vessel with the adsorbent to the environment, where it can produce a phase transition, i.e. the melting of ice, naphthalene, diphenyl ether, phenol, diphenylmethane, etc. A classical example of an isothermal calorimeter is the Bunsen calorimeter the operation of which is based on the measurement of the change of volume of a substance part of which melts under the action of the heat of adsorption released.

In the last group calorimeters considered, known as diathermal, the transfer of heat from the vessel containing adsorbent is not prevented but its rate is controlled to a certain level. The total amount of energy released in the form of heat is calculated from the instantaneous thermal balance. The rate of heat transfer may be relatively large if in the jacket we have a gas under a given pressure, while in the case of a vacuum it may be considerably reduced.

To conclude, we wish to make some general comments concerning all these types of calorimeter. A given quantity of energy released in the form of heat at any point of the calorimeter should give the same response of the measuring instrument. The measurement of temperature in the adsorbent layer also presents an important problem (it is advantageous to take measurements in parallel at various sites of the layer), as does simultaneous and uniform supply of the adsorbate to the whole surface area of the adsorbent. In simple designs of adsorption calorimeters some of these requirements are only partially fulfilled.

Calorimetric techniques find, owing to their continuous development and automation, ever-wider applications not only in adsorption studies but also in explaining many problems connected with adsorption on the surface of active carbons and other adsorbents.

References

[1] Avgul', N. N., Berezin, G. I., Kiselev, A. V., Lygina, I. A., *Zh. Fiz. Khim.* **30**, 2106 (1956).
[2] Paryjczak, T., *Gas Chromatography in Adsorption and Catalysis*, PWN — Ellis Horwood, Warsaw-Chichester 1986.

References

[3] Stoeckli, H. F., Kraehenbuehl, F., *Carbon* **19**, 353 (1981).
[4] Stoeckli, H. F., Kraehenbuehl, F., *ibid.* **22**, 297 (1984).
[5] Stoeckli, H. F., *Proceedings of the 14th International Carbon Conference, CARBON'86*, Baden-Baden 1986, p. 271.
[6] Smíšek, M., Černý, S., *Active Carbon*, Elsevier, Amsterdam-London-New York 1970.
[7] Young D. M., Crowell, A. D., *Physical Adsorption of Gases*, Butterworths, London 1962.
[8] Gregg, S. J., Sing, K. S. W., *Adsorption Surface Area and Porosity*, Academic Press, London-New York 1967.
[9] Kiselev, A. V., Dreving, V. P. (Eds.), *Eksperimentalnye metody v adsorbtsii i molekularnoi khromatografii* (Experimental Methods in Adsorption and Molecular Chromatography), Izd. Mosk. Univ., Moscow 1973.

CHAPTER 7

Adsorption from the Liquid Phase

7.1 FUNDAMENTAL RELATIONSHIPS FOR THE SURFACE LAYER

The molecules in the surface layer at the interface are exposed to the action of molecules situated in both phases. By the surface layer we understand those parts of each of the phases in which the average energy state of the molecules, resulting from intermolecular interactions, is different from that in the bulk of each phase.

The surface area of the interface may undergo expansion, when the geometric properties of the system are changed, however, this always involves the supply of energy to the system. The differential of this energy consists of two contributions: (i) of the work of the volume change dA_{vol} and (ii) of the work of expansion of the surface area of the interface dA_{surf}. We recall here the well-known equation for the work that must be supplied to the system to enable expension of the interfacial surface area:

$$dA = dA_{vol} + dA_{surf}. \tag{7.1}$$

Since

$$dA_{vol} = -pdV \quad \text{and} \quad dA_{surf} = \sigma dq \tag{7.2}$$

where p is the constant pressure against which the system performs work, dV is the differential of the volume increase of the system, dq is the surface area by which the interface increases, and σ is the proportionality coefficient referred to as the surface tension. Substituting equations (7.2) into (7.1) we obtain:

$$dA = -pdV + \sigma dq. \tag{7.3}$$

Equation (7.3) can be regarded as the definition of the surface tension at the interface expressed in force per unit length ($N\ m^{-1}$) or in work per unit surface area ($J\ m^{-2}$). By definition, the surface tension is the work necessary to expand the surface area of the interface, and $\sigma > 0$.

If we consider the surface layer as an open system, then the equation for the differential of the internal energy (dU) of the system has the form:

$$dU = dQ + dA + \Sigma G_i dn_i \tag{7.4}$$

where $\Sigma G_i dn_i$ accounts for the exchange of molecules between the system and the environment, and dQ and dA are the energies exchanged by the system with the environment in the form of heat and work, respectively.

By definition, the thermodynamic potential G of the system is:

$$G = H - TS \tag{7.5}$$

where H is the enthalpy, S is the entropy and T is the temperature in K.

Differentiating equation (7.5) we obtain:

$$dG = dH - TdS - SdT. \tag{7.6}$$

By definition:

$$H = U + pV \tag{7.7}$$

and after differentiation

$$dH = dU + pdV + Vdp. \tag{7.8}$$

If we substitute equations (7.3), (7.4) and (7.8) into (7.6) and reduce similar terms (noting that $dQ = TdS$) we get:

$$dG = Vdp - SdT + \sigma\, dq + \Sigma G_i dn_i \tag{7.9}$$

and hence

$$\sigma = \left(\frac{\partial G}{\partial q}\right)_{p,T,n} \tag{7.10}$$

This equation, like equation (7.3), is considered to be the definition of surface tension.

If both phases consist of the same chemical species, for instance the pure liquid and its saturated vapour, then the interface differs from the bulk phases solely as regards structure (the distribution of molecules in space and the potential energy). When there is more than one component in the system, the surface layer interface usually differs as regards concentration (the ratio of the numbers of molecules of the components) from each of the two phases. This change of concentration of a substance at the interface is known as adsorption. As regards the varieties of contacting phases we can consider the adsorption process, apart from the case discussed in the preceding chapters, as taking place in the following systems: liquid–gas, solid–liquid and liquid–liquid.

7.2 GIBBS ADSORPTION ISOTHERM

The process of adsorption at the liquid–gas interface was described in detail in thermodynamic terms by Gibbs in 1878. His approach continues to be accepted and is the basis of contemporary theory of surface phenomena.

Here we concentrate on finding a relation between adsorption at the solution–gas interface and the concentration of the solution[†]. Let us assume that we have a two-component and two-phase system. If we apply to this system equation (7.9) we obtain:

$$dG = Vdp - SdT + \sigma dq + G_{1(l)} dn_{1(l)} + G_{2(l)} dn_{2(l)} +$$
$$+ G_{1(g)} dn_{1(g)} + G_{2(g)} dn_{2(g)} \qquad (7.11)$$

where: $G_{1(l)}, n_{1(l)}, G_{2(l)}, n_{2(l)}, G_{1(g)}, n_{1(g)}, G_{2(g)}, n_{2(g)}$ are the partial molar thermodynamic (chemical) potentials and numbers of moles of components 1 and 2 in the liquid and gas phases.

Assume that in the system, adsorption of component 2 takes place. At constant temperature and pressure this phenomenon is accompanied by a change of concentration of component 2. By differentiating equation (7.10) with respect to the number of moles of component 2 we obtain:

$$\frac{\partial \sigma}{\partial n_{2(l)}} = \frac{\partial^2 G}{\partial q \partial n_{2(l)}}, \qquad \frac{\partial \sigma}{\partial n_{2(g)}} = \frac{\partial^2 G}{\partial q \partial n_{2(g)}} \qquad (7.12)$$

Since by definition:

$$G_{2(l)} = \frac{\partial G}{\partial n_{2(l)}} \quad \text{and} \quad G_{2(g)} = \frac{\partial G}{\partial n_{2(g)}} \qquad (7.13)$$

[†] Here this relation is derived following Tomassi [1] and not in the original form given by Gibbs.

then substituting equation (7.13) into (7.12) we obtain:

$$\frac{\partial \sigma}{\partial n_{2(l)}} = \frac{\partial G_{2(l)}}{\partial q} \quad \text{and} \quad \frac{\partial \sigma}{\partial n_{2(g)}} = \frac{\partial G_{2(g)}}{\partial q} \tag{7.14}$$

At equilibrium we have:

$$G_{2(l)} = G_{2(g)} \tag{7.15}$$

and hence

$$\frac{\partial \sigma}{\partial n_{2(l)}} = \frac{\partial \sigma}{\partial n_{2(g)}} = \frac{\partial \sigma}{\partial n_2} \tag{7.16}$$

In the case of positive adsorption in a liquid solution, when the concentration of component 2 is increasing in the surface layer and decreasing in the bulk of the solution, we obtain:

$$\frac{\partial G_{2(g)}}{\partial q} < 0 \quad \text{and} \quad \frac{\partial \sigma}{\partial n_2} < 0 \tag{7.17}$$

if $dq > 0$ and $dG_2 < 0$.

The physical sense of equation (7.17) in the case of positive adsorption on the surface of a liquid solution is as follows: the increase of the number of moles of component 2 with respect to component 1 produces a lowering of the surface tension. In the case considered, component 2 is referred to as a surfactant.

In the case of negative adsorption the surface tension increases with the increase of concentration of the component adsorbed negatively from the given phase. Here component 2 is surface-inactive.

Since equation (7.15) is valid, the equations analogous to (7.17) hold for component 2 in the gas phase. The adsorption of a particular component is negative or positive simultaneously in both phases.

The effects of negative adsorption are comparatively small, while those of positive adsorption are very pronounced. Negative adsorption is observed quite often, for instance in the case of substances that dissociate into ions, while positive adsorption occurs in aqueous solutions, e.g. of alcohols.

Assume that in preserving the state of equilibrium, we change the surface area of the interface and the amount of component 2 in the solutions, then:

$$G_{2(l)} = \varphi(q, n_2) \tag{7.18}$$

Writing function (7.18) in the form of a total differential and accounting for the equilibrium condition we obtain:

$$dG_{2(l)} = -\frac{\partial G_{2(l)}}{\partial q} dq + \frac{\partial G_{2(l)}}{\partial n_{2(l)}} dn_{2(l)} = 0 \qquad (7.19)$$

Hence

$$\frac{\partial G_{2(l)}}{\partial q} dq = -\frac{\partial G_{2(l)}}{\partial n_{2(l)}} dn_{2(l)}$$

and

$$\frac{\partial G_{2(l)}}{\partial q} \cdot \frac{\partial n_{2(l)}}{\partial G_{2(l)}} = -\left(\frac{\partial n_{2(l)}}{\partial q}\right)_{G_{2(l)}}$$

while considering (7.15) we obtain

$$\left(\frac{\partial n_{2(l)}}{\partial q}\right)_{G_{2(l)}} = \left(\frac{\partial n_{2(l)}}{\partial q}\right)_{G_2} = \Gamma_{surf\,2} \qquad (7.20)$$

where $\Gamma_{surf\,2}$ is called the surface concentration of component 2.

If we take account of (7.14), the surface concentration $\Gamma_{surf\,2}$ can be expressed as follows:

$$-\Gamma_{surf\,2} = \frac{\partial \sigma}{\partial n_2} \frac{\partial n_2}{\partial G_2} = \left(\frac{\partial \sigma}{\partial G_2}\right)_q \qquad (7.21)$$

If both phases are ideal solutions, we can assume that the following equations are true:

$$G_{2(l)} = G_{02(l)} + RT \ln c_2$$
$$G_{2(g)} = G_{02(g)} + RT \ln p_2 \qquad (7.22)$$

where: $G_{02(l)}$ and $G_{02(g)}$ are constant.

By differentiating equations (7.22) with respect to the surface tension we obtain:

$$\frac{\partial G_{2(l)}}{\partial \sigma} = RT \frac{\partial \ln c_{2(l)}}{\partial \sigma}$$

$$\frac{\partial G_{2(g)}}{\partial \sigma} = RT \frac{\partial \ln p_2}{\partial \sigma} \qquad (7.23)$$

On comparing equations (7.23) and (7.21) we find:

$$-\Gamma_{surf\,2} = \frac{1}{RT}\frac{\partial\sigma}{\ln c_2}$$

$$-\Gamma_{surf\,2} = \frac{1}{RT}\frac{\partial\sigma}{\ln p_2} \tag{7.24}$$

Equations (7.24) are known as the Gibbs adsorption isotherms. If the relationship $\sigma = f(\log c_2)$ and $\sigma = \varphi(\log p_2)$ are found experimentally, $\Gamma_{surf\,2}$ is determined by differentiating the relevant curves. Note that the Gibbs adsorption isotherms have been verified fully by experiment.

The quantity $\Gamma_{surf\,i}$ considered by Gibbs, known as the surface excess of component i in the surface layer, the Gibbs absolute adsorption or, in brief, the Gibbs adsorption, means physically the difference in content of component i in the surface layer (per unit surface area) and the corresponding volume of the bulk phase.

7.3 ISOTHERM FOR ADSORPTION BY A SOLID FROM BINARY LIQUID SOLUTIONS

The mechanism of adsorption from liquid solutions differs significantly from that of adsorption from a single-component gaseous phase as discussed in detail in preceding chapters. The components adsorbed from the liquid solution form a tight layer on the adsorbent surface. The equilibrium of the adsorption process is considered as the equilibrium between two solutions of different compositions: the surface solution and the bulk solution. A change of concentration of the bulk solution leads to a change of composition of the surface layer, i.e., it causes mutual displacement of the solution components from the surface layer. In the surface solution (surface layer) no free sites appear, but the molecules of one component merely replace those of the other.

Adsorption from the liquid phase is determined either from the difference of compositions of the bulk and surface solutions or from the composition of the surface layer. In the former case we are concerned with composite adsorption (apparent or excess), and in the latter with real or individual adsorption. The physical meaning of excess adsorption and the definition of the magnitude of adsorption have been given in the preceding paragraph for a liquid two-phase and two-component system when adsorption occurs at the solid–liquid interface.

In the rather common situation of adsorption from a two-component liquid solution onto a solid (e.g. active carbon) the miscibility of the solution components may be infinite or limited. Let us consider each case separately.

The first occurs when we are concerned with two-component solutions of numerous organic solvents (e.g. hydrocarbons, their chloro-derivatives, alcohols, esters, amines, etc.). If by n^0 we denote the total number of moles of the solution components ($n^0 = n_1^0 + n_2^0$, where n_1^0 and n_2^0 are the numbers of moles of the solution components 1 and 2, respectively) and by m the mass of the adsorbent used, then if x_1^0 and x_1 are the mole fractions of component 1 in the initial solution and at adsorptive equilibrium, then the excess adsorption may be written in the form:

$$n_1^{\sigma(n)} = \frac{n^0(x_1^0 - x_1)}{m} \qquad (7.25)$$

and for 1 g of adsorbent:

$$n_1^{\sigma(n)} = n^0(x_1^0 - x_1) = n^0 \Delta x_1 = n_1^s - (n_1^s + n_2^s) x_1 \qquad (7.26)$$

where n_1^s and n_2^s are the numbers of moles of components 1 and 2, respectively, in the surface solution. The above relationship, which represents the isotherm of excess adsorption, is also known as the composite adsorption isotherm [2]. Since from investigation of adsorption from solutions we usually obtain excess adsorption isotherms, the classification of these is important. Often their shape

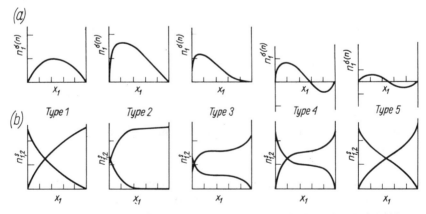

Fig. 7.1 Classification of the isotherms of excess (a) and real adsorption (b) from binary solutions according to Nagy and Schay [3, 4].

provides a criterion for such a classification. Nagy and Schay [3, 4] proposed in 1960 a classification of adsorption from binary solutions of infinite miscibility (see Fig. 7.1). For each of the five types of excess adsorption, the individual isotherms have also been given. In the case of adsorption on active carbons, as for other adsorbents, we encounter every type of isotherm in this classification depending on the properties of the adsorbent surface and the nature of the organic solvents comprising the binary solution. In Fig. 7.2 the adsorption isotherms on active carbon are shown, by way of example, of n-butylamine and methyl acetate from benzene solution.

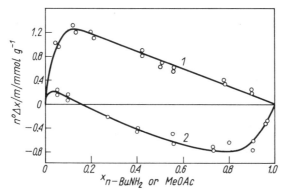

Fig. 7.2 Isotherms of excess adsorption on active carbon from coconut shells of: 1 — n-butylamine and 2 — methyl acetate from benzene solution (after Blackburn, Kipling and Tester [5]).

As regards adsorption from binary liquid solutions of limited miscibility, solutions of solid substances in liquid solvents provide a typical example. Giles et al. proposed in 1960 a method of classifying the adsorption isotherms for such systems [6] — see Fig. 7.3. In this classification four main classes of adsorption isotherm are distinguished according to their initial behaviour. Class S curves are obtained when the solvent is strongly adsorbed, strong intermolecular attraction occurs in the adsorption layer, and the adsorbate molecules are monofunctional. A vertical or possibly skew arrangement of the adsorbed molecules is observed here. Class L (Langmuir-type) refers to cases when there is no strong competition on the part of the solvent molecules in the coverage of the active sites on the adsorbent surface by the adsorbate. Here the adsorbate molecules are usually arranged horizontally in the adsorption layer. Curves of type H occur when the

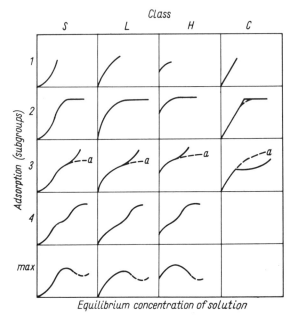

Fig. 7.3 Classification of the adsorption isotherms of solids from their solutions according to Giles et al. [6].

affinity between the adsorbate and adsorbent is very high even if the solutions are very dilute. This may be due, for instance, to the occurrence of chemisorption or adsorption of polymer molecules or ionic micelles. Class C curves refer to the case of constant partition of the adsorbed substance between the surface layer and the bulk solution, which occurs for some fibrous adsorbents.

Within the particular classes, subgroups 1, 2, 3, 4 as well as a group denoted as 'max' (i.e. showing a maximum) are distinguished depending on the shape of the isotherms at higher equilibrium concentrations of the adsorbate in the solution. As an example of adsorption isotherms belonging to various groups within one class of the Giles classification, one can mention the adsorption isotherms on active carbon of iodine from alcohol (e.g. methanol) solutions [7]. If an electrolyte (sodium iodide) is added to the solution before it is brought into contact with the adsorbent, curves belonging in the absence of electrolyte to group 2 of class L change their shape to that of group 4 of the same class. This effect is considered in the case to be due to the equilibrium:

$$I_2 + I^- \rightleftharpoons I_3^-.$$

In this reaction the equilibrium constant in methanol solution is relatively high and its logarithm is equal to 4.2 at 25°C. Iodine is adsorbed on the surface of active carbon only in molecular form and only to a very small degree in ionic form. Thus, as long as the iodine concentration in solution is smaller than that of the electrolyte, iodine occurs as I_3^- and hence we observe a distinct lowering of its adsorption. It is only at equilibrium concentrations (total of I_2 and I_3^-) greater than that of the electrolyte that the quantity of free iodine increases with the increase of its total amount in solution, producing a rapid growth of adsorption (hence the wavy shape of the curve). The changes in shape of the adsorption isotherm are illustrated in Fig. 7.4.

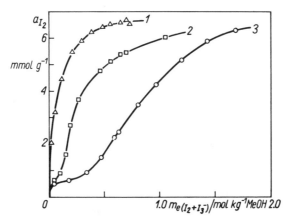

Fig. 7.4 Adsorption isotherms on CWZ-3 active carbon of iodine dissolved in: 1 — methanol, 2 — methanolic NaI (0.15 mol dm^{-3}), 3 — methanolic NaI (0.5 mol dm^{-3}) (after Świątkowski [7]).

The general approach considered here to adsorption from binary liquid solutions of limited miscibility is very important. The series of contributions of Manes et al. [8–13] bring a wealth of new data on the subject. These authors have been considering the possibility of applying Polanyi's potential theory of adsorption for describing adsorption equilibria in systems composed of active carbon and aqueous or non-aqueous solutions of several dozen adsorbates of varying properties (including solids and liquids with different solubilities in the solvents used).

Recent developments regarding the discription of iodine adsorption from solutions in various organic solvents are given in refs. [14, 15].

7.4 ADSORPTION ON ACTIVE CARBONS

Most experimental studies of adsorption from solutions concentrate on determination of the adsorption isotherms of the individual substances from dilute aqueous solutions on carbon adsorbents. Derylo-Marczewska and Jaroniec [16] give many experimental data on the adsorption of organic substances from a wide range of solvents, particular attention being paid to adsorption from aqueous solutions.

The influence of the properties of the components of adsorption systems and the experimental conditions of adsorption equilibrium have been the object of many studies. First, adsorption was considered as a direct function of the properties of the organic solute such as molecular mass, geometry and size, its functional groups, polarity, polarizability and solubility. Al-Bahrani and Martin [17] investigated the effect of the solute molecular structure on adsorption by examining the experimental sorption data of different benzene and phenol derivatives on activated carbon. These authors have stated that the increase of adsorptive capacity with increasing molecular mass is a general tendency resulting from the greater affinity of larger molecules to the carbon surface. They also studied the effect of different functional groups on adsorption, noting the role of hydrogen bonding. However, these authors did not find any general correlation between the adsorptive capacity and solubility or dipole moment. Rovinskaya and Koganovskii [18, 19] have also investigated the dependence of adsorption on the solute structure. The effects of polarity, solubility and molecular mass on the adsorption of various organic solutes from water was studied by Akhmadeev et al. [20], who found that simple correlations exist between these factors and the adsorption capacity. The quoted authors also investigated the differences in the adsorption of homologous alcohols, and concluded that an increase in molecular mass (and hence the extension of the hydrocarbon chain) causes an increase in the adsorption capacity. El-Dib and Badawy [21] studied the adsorption of benzene, toluene, o-xylene, and ethylbenzene on active carbon, and applied the experimental data to the Freundlich equation. Comparing the parameters of the Freundlich isotherm for all the solutes investigated, they estabilished their dependence on the chemical structure and solubility. Abe et al. [22] carried out extensive investigations on correlations between the adsorbabilities of a large number of organic compounds and many different physical parameters such as molecular mass, number of carbon atoms in the molecule, molar volume, solubility and parachor.

These authors found that a linear relationship exists between the logarithm of the adsorbability and some of these parameters, enabling adsorption to be predicted.

The adsorption process is also affected by the nature of the adsorbent, i.e., its surface area, porous structure, and surface properties. The great structural and surface complexity of the active carbons is clear from the variety of adsorption mechanism which may be classified into three types: physical, chemical or electrostatic ones [16]. Exhaustive reviews of carbon structures and adsorption properties have been given [23–25]. Puri *et al.* [26] investigated the adsorption mechanism of hydroquinone on different types of carbon black. The amount of hydroquinone physically adsorbed is a function of the specific surface area and does not depend on the surface properties of the adsorbent. However, in the case of chemically adsorbed hydroquinone, the nature of the surface was observed to play a significant role. Elsewhere, Puri [27] studied the effect of modification of the carbon surface by different treatments on the shape of the adsorption isotherms of some organic substances. In studying the adsorption of nitro compounds on four modified carbons, Glushchenko *et al.* [28] found a decrease of the amount adsorbed after oxidation. This confirmed the conclusion that adsorption occurs mainly on that part of the surface which does not have functional groups containing oxygen. The influence of carbon surface acidity was also widely investigated by Oda and Yokokawa [29] for the adsorption of variety of organic substances on different activated carbons. Elsewhere [30] comparison is made between the mechanisms of adsorption of benzoic acid and phenol from aqueous solution. In the case of benzoic acid a correlation between adsorption and surface acidity was found to exist. The amount of phenol adsorbed increased with the decrease of surface acidity achieved after heat-treatment. Coughlin and Ezra [31] reported that adsorption of phenol and nitrobenzene decreases with increased levels of surface oxides, but their proposal of a two-step adsorption of phenol seems doubtful.

The effectiveness of adsorptive purification methods requires that the optimal conditions of the sorption process be determined. To this end, not only the selection of the adsorbent but also the determination of experimental conditions such as solution pH and ionic strength or temperature is very important. The effect of temperature on the course of the adsorption process has been studied by many authors [32–34], most reporting that solute adsorption decreases with increasing temperature.

The effect of pH on adsorption of weak organic electrolytes has been investigated extensively [32, 35–37] both experimentally and theoretically. These studies reveal the importance of changes of the carbon surface charge with pH for the adsorption equilibrium. The pH effect was considered to be a complex phenomenon that changes the nature of the solid surface, of the adsorbate molecules, the solution composition, and the ionic strength due to the introduction of additional components.

The mathematical description of experimental adsorption data is of paramount importance. The theoretically-based Langmuir isotherm gives a good description of only a few experimental systems. However, more advanced theoretical approaches have been proposed recently. The lack of a theoretical isotherm fitting to the experimental data over a wide range of concentrations has resulted in the use of various empirical equations.

The Freundlich isotherm, first proposed for adsorption of a single gas, was adapted to adsorption from dilute solution by the simple replacement of the gas pressure by the solute concentration:

$$a_i = k(c_i)^m \tag{7.27}$$

where a_i is the quantity of solute i adsorbed, c_i is its concentration and k and m are constants. This equation was used successfully to analyse many experimental data [17, 21, 38, 39]. However, it should be noted that the Freundlich isotherm is thermodynamically inconsistent: at low concentrations it does not fulfil Henry's law and at high concentrations it tends to infinity. These features of the Freundlich isotherm lead to great discrepancies between its predictions and experimental data over ranges of very low and high concentrations, though agreement at intermediate concentrations may be quite good. Thus the usefulness of this isotherm is restricted to a relatively narrow concentration range.

Another popular experimental isotherm was proposed by Radke and Prausnitz [40]:

$$1/a_i = 1/(\alpha c_i) + 1/[\beta (c_i)^m] \tag{7.28}$$

where α, β and m are constants. This equation was successfully used for the description of several experimental isotherms [41, 42]. At low concentrations it reduces to Henry's law, while at high concentrations it takes the form of the Freundlich isotherm. Again, if m tends to zero then the Langmuir equation is obtained.

Jossens et al. [43] adapted the three-parameter Newman equation to adsorption from dilute solution:

$$a_i = [(\alpha c_i) + m(\beta c_i^m)^{-1}] \left[(\alpha c_i)^{-1} + (\beta c_i^m)^{-1} \right]^{-2} \qquad (7.29)$$

This equation is almost equivalent to the Radke–Prausnitz equation (7.28).

An analogue of the Harkins–Jura equation was proposed by Iyer and Wariyar (see Brown and Everett [44]) for analysis of adsorption from dilute solution:

$$\ln c_i = \tilde{\alpha} - \tilde{\beta}/(a_i)^2 \qquad (7.30)$$

where $\tilde{\alpha}$ and $\tilde{\beta}$ are constants.

John et al. [45] modified John's isotherm for single-solute adsorption to:

$$\log \{\log [(c_i/c_{s,i}) \cdot 10^{N^*}]\} = C^* + D^* \log a_i \qquad (7.31)$$

where: $c_{s,i}$ is the solubility of the ith solute, N^* is an integer greater than unity and C^* and D^* are constants.

One possible approach leading to the description of adsorption equilibria in the system, solid adsorbent—dilute aqueous solution of an organic substance, is the application of the theory of volume filling of micropores [46–49]. It has been shown that the D–R equation may be used successfully here if we replace the pressure ratio p_0/p by the concentration ratio c_0/c (where c_0 is the concentration of the saturated solution and c is the adsorbate concentration at equilibrium). Let us now discuss this approach using examples from the literature.

Korta [46] presented experimental data obtained for five different active carbons in the form of common plots in the $\log a$–$(\log p_0/p)^2$ and $\log a$–$(\log c_0/c)^2$ coordinate systems for vapour adsorption and adsorption from solution, respectively — see Fig. 7.5. The experimental plots lie along straight lines for the microporous carbon used. The values W_0 calculated on the basis of the plot showed very good agreement (the differences between the values W_0 for different adsorbents do not exceed 2.5%).

Deviations from rectilinearity of the D–R isotherm occurred for carbons with greater pores. Adoption of the equation for carbons of the second structural type, i.e. in the $\log a$–$\log p_0/p$ or $\log a$–$\log c_0/c$ coordinate systems, yielded a linear plot of the experimental points.

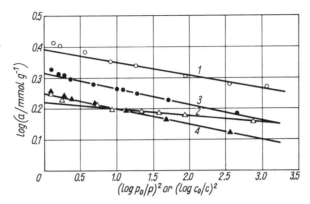

Fig. 7.5 Adsorption isotherms: 1 — benzene, 2 — n-hexane vapours, 3 — m-cresol, 4 — caproic acid from aqueous solution onto gas carbon H obtained from coke (with micropores dominant), according to the D–R equation (after Korta [46]).

Koganovskii and Levchenko [47] studied adsorption from solutions of many organic substances on active carbon KAD. The results are shown in the form of $\log(aV)$—$(\log c_0/c)^2$ plots for adsorption from aqueous solution in Fig. 7.6. All the experimental points lie along straight lines which intersect the ordinate at one point ($\log(aV) = \log W_0 \approx -0.4$) what indicates that the D–R equation is fulfilled for each of the adsorbates used.

In the case of molecules with either a large dipole moment (p-nitroaniline) or an electric charge (aniline cations), mutual repulsion

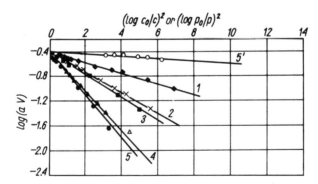

Fig. 7.6 Adsorption isotherms: 1 — chloroform, 2 — p-chloroaniline, 3 — nitrobenzene, 4 — epichlorohydrin, 5 — benzene from aqueous solution, 5' — benzene vapour on active carbon KAD according to the D–R equation (after Koganovskii and Levchenko [47]).

of the adsorbate molecules takes place in the adsorption layer, which leads to deviations from the rules governing the theory of volume filling of micropores (log $W_0 < -0.4$). A part of the adsorption space is in this case occupied by solvent molecules. In the course of the adsorption of aniline, the deviation observed is due to protonation of aniline by water molecules which leads to the creation of strongly polar complexes whose mutual repulsion inhibits filling of the micropores. It is only at high equilibrium aniline concentrations that the quantity of adsorbed water molecules decreases, and hence protonation is possible to a much smaller degree. When the adsorption space is completely filled with aniline molecules, the effect described disappears altogether and at low W_0 takes the value of ca. -0.4 (as in the case of adsorption of benzene vapour). The deviations may also be due to other reasons. The authors quoted give as illustration of this the adsorption of alcohols from aqueous solution. The adsorption isotherms of propyl, butyl, and hexyl alcohol expressed as plots of $\log (aV)$ versus $(\log c_0/c)^2$ are indeed linear, but the log W_0 value given by the intersection of these straight lines on the ordinate is much smaller (about -0.6). One may surmise that even on maximum filling of the adsorption space with the adsorbate molecules, water molecules hydrogen-bonded to the alcohol molecules are included. The results of Koganovskii and Levchenko [47] lead us to the conclusion that the D–R equation may be applied to systems in which no interaction occurs between the adsorbate molecules or where this interaction changes insignificantly as the adsorption space is filled up or where the adsorbate molecules do not hydrogen-bond with water.

These authors devoted considerable attention to the similarity coefficient β. They found that in order to convert the adsorption isotherm of a given solute into that of benzene vapour, one can employ a similarity coefficient equal to the product of the coefficient β for converting the adsorption isotherm of this substance into that of benzene from an aqueous solution, and the coefficient β for converting the adsorption isotherm of benzene from aqueous solution into that of benzene vapour. This finding is confirmed by the results of calculations based on experimental data.

The problem of adapting the theory of volume filling of micropores to adsorption from solution was also tackled by Stadnik and Él'tekov [48]. These authors presented the adsorption isotherms of toluene from aqueous solution on four different active carbons as plots of $\log a$ versus $(\log c_0/c)^2$. They proved mathematically that the form of the D–R equation is linear when the term $\log c_0/c$ is squared.

Further efforts devoted to these problems had the aim of explaining the role of various factors affecting the process of adsorption of organic compounds from solutions onto active carbons.

Levchenko, Kakulina and Koganovskii [50] advocated p-chloroaniline as a standard adsorbate in probing the capillary structure of carbon adsorbents. These authors found from their analysis of its adsorption isotherms from aqueous solutions onto carbon KAD and anthracite products subjected to activation (to different degrees of burn-off) that exclusive adsorption of p-chloroaniline takes place only in the smallest micropores. Micropores of greater radii and mesopores are filled both with adsorbate and solvent. Making use of the de Boer t-method these authors determined the proportion of the pores filled solely with p-chloroaniline relative to the total volume of micropores as determined from the D–R equation.

The effect of the structure of the adsorbate molecules on the limiting volume determined for the adsorption space was studied by Koganovskii and Sal'kova [51]. These authors found that in the case of adsorption of benzene derivatives (ethyl-, m-diethylbenzene, styrene, m-divinylbenzene, and m-ethylstyrene) from aqueous solution on carbon KAD, the experimental data plotted as $\log(aV)$ versus $(\log c_0/c)^2$ show deviations from linearity at high equilibrium concentrations, while when plotted as $\log(Va)$ versus $\log c_0/c$, the isotherms are linear. The values of W_0 determined graphically from the plots differ considerably both for various adsorbates and even for a single adsorbate depending on the coordinate system in which the experimental data have been presented. Evidently the W_0 values are strongly dependent on the structure of the adsorbate molecules.

In further studies, Stadnik and Él'tekov [52] found that the D–R equation satisfactorily describes the adsorption of organic substances from aqueous solution only above a certain limiting equilibrium concentration (denoted as c^*) below which the experimental data are much better described by Henry's law in the form:

$$a = a_0 Kc \qquad (7.32)$$

where K is Henry's constant and a_0 is the maximum adsorption. These authors have derived an equation defining c^* that limits the applicability of the D–R equation, as well as the equation describing the dependence of K on the B constant in the D–R equation. For adsorption isotherms of benzene from aqueous solution on different active carbons, these authors have obtained linear relationships (Fig. 7.7) for concentrations

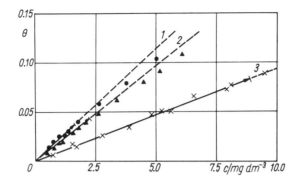

Fig. 7.7 The initial linear sections of the adsorption isotherms of benzene from aqueous solution on active carbons: 1 — BAU, 2 — SKT, 3 — AG-3 (the solid lines correspond to the region followed by Henry's law) (after Stadnik and Él'tekov [52].

lower than c^* on plotting the data points as θ versus c (where $\theta = a/a_0$ is the degree of filling of the adsorption space).

Stadnik and Él'tekov [53] have used the dependence of K on B to relate the thermodynamic quantities characterizing the process of adsorption from solution to the capillary structure of the adsorbent and properties of the adsorbate.

These authors have used their equations to calculate the changes of the molar thermodynamic potential (the molar Gibbs free energy) for adsorption of benzene and several of its derivatives from aqueous solution on carbon KAD and compared these changes with those found experimentally. Very good agreement was obtained (the deviations did not exceed 7%). The procedure proposed by Stadnik and Él'tekov makes it possible to determine the thermodynamic characteristics of the adsorption process from knowledge of the capillary structure of the active carbon (constant B in the D–R equation) and of the properties of the adsorbate (β, c_0, V). The value of the structural energetic constant B is usually determined experimentally (by determination of the adsorption isotherm followed by calculations based on the TVFM).

Stadnik and Él'tekov [54] have also developed a theoretical method of calculating the constant B in the D–R equation for adsorption of organic vapours. They obtained good agreement between the values of this constant calculated and determined experimentally. These authors also applied this method to the adsorption of organic substances from aqueous solution [55]. A method has also been given of converting the constant B on passing from adsorption of a particular organic substance from the gas phase to adsorption of the same substance from solution. In order to check this method Stadnik and Él'tekov calculated the values of B for two different active carbons and toluene as adsorbate (used in [50]) and also for carbon KAD and benzene (used in [48]). As for adsorption from the gas phase [54] the authors obtained good agreement between their experimental results and those calculated on the basis of the proposed equations.

Based on these various considerations, and on the kinetics and dynamics of adsorption of organic substances from aqueous solution, Stadnik and Él'tekov [56] developed a mathematical model of the process. This model is claimed to enable, for a small number of experimental data, calculations of sorption devices for water purification and to select, for particular conditions, the most effective adsorbent.

Fairly recently Jaroniec and Choma [57, 58] have adapted to the case of adsorption from dilute solution, the isotherm proposed for the adsorption of gases on highly heterogeneous microporous solids (see Chapter 5). From the survey of literature given above, it is clear that most publications devoted to adsorption from dilute aqueous solution onto active carbons, concentrate on the determination of the sorptive capacity of these adsorbents. Jaroniec and Choma suggest that their method, apart from determining the adsorptive capacity of adsorbents, also extends to evaluation of the adsorption potential distribution function and the structural parameter B distribution function (see Chapter 5). These distributions determined from adsorption data characterize the structural and energetic heterogeneity of the microporous active carbons. The heterogeneity both of the active carbon microporous structure and its surface play an important role in adsorption from dilute aqueous solution.

The adsorption isotherm proposed by Jaroniec and Choma [58] for describing adsorption from dilute aqueous solution has the following form:

$$\theta_t(A) = \left(\frac{q^*}{q^* + A^2}\right)^{n+1} \tag{7.33}$$

where

$$\theta_t(A) = a/a_{mi}^0 \quad \text{and} \quad q^* = q\beta^2 \tag{7.34}$$

and

$$A = RT \ln(c_0/c) \tag{7.35}$$

(A is the adsorption potential).

In equations (7.33)–(7.35) a is the equilibrium adsorption, a_{mi}^0 is the adsorption capacity of the micropores, q and m are the parameters of the micropore distribution function which is a gamma distribution function, $\theta_t(A)$ is the degree of filling of the micropore adsorption space, β is the similarity coefficient characterizing the adsorbate, R is the universal gas constant, T is the absolute temperature, c is the concentration of the dissolved substance and c_0 is the concentration of the dissolved substance at saturation.

A direct relationship occurs between the adsorption isotherm equation (7.33) and the adsorbent micropore distribution function with respect to the structural parameter B described by the relationship:

$$F(B) = [q^{n+1}/\Gamma(n+1)] B^n \exp(-qB) \qquad (7.36)$$

and the adsorption potential distribution function:

$$X(A) = 2(n+1)(q^*)^{n+1} A (q^* + A^2)^{-(n+2)}. \qquad (7.37)$$

These functions, determined from the adsorption isotherms of phenol, p-nitrophenol and benzoic acid are given by way of example in Figs. 7.8 and 7.9. These functions provide a detailed specification of the test adsorbent.

These authors [57, 58] claim that equation (7.33) when used for adsorption from solution proves very useful as regards the description of adsorption on active carbons of organic substances from dilute aqueous solution.

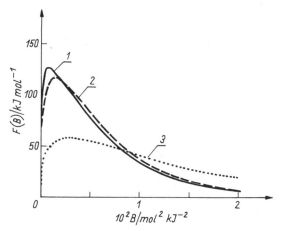

Fig. 7.8 The gamma distribution functions $F(B)$ calculated from equation (7.36) for active carbon BIO I: 1 — phenol, 2 — p-nitrophenol, 3 — benzoic acid.

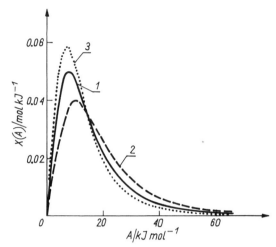

Fig. 7.9 The distribution functions $X(A)$ calculated from equation (7.37) for active carbon BIO I: 1 — phenol, 2 — p-nitrophenol, 3 — benzoic acid.

Though the process of adsorption on active carbons from the liquid phase has been much less investigated quantitatively than that from the gas phase and is less mechanistically defined, adsorption from dilute solution has found wide practical application, and hence the literature on this subject is quite exhaustive [46, 59–62]. However, most works deal with the practical aspects of the process. Only in recent years have more theoretically-based works appeared. These problems — both theoretical and practical — have been highlighted in this book rather arbitrarily and necessarily very briefly. It is hoped, however, that the references cited will enable the reader to gain access to this developing subject.

References

[1] Tomassi, W., *Termodynamika chemiczna (Chemical Thermodynamics)*, vol. 3, PWN, Warsaw 1955.
[2] Kipling, J. J., *Adsorption from Solutions of Non-Electrolytes*, Academic Press, London-New York 1965.
[3] Nagy, L. G., Schay, G., *Magyar Kem. Folyoirat* **66**, 31 (1960).
[4] Schay, G., Nagy, L. G., Shekrenyesy, T., *Periodica Polytech.* **4**, 95 (1960).
[5] Blackburn, A., Kipling, J. J., Tester, D. A., *J. Chem. Soc.* 2373 (1957).
[6] Giles, C. H., MacEvan, T. H., Nakhwa, S. N. et al., *J. Chem. Soc.* 3973 (1960).
[7] Świątkowski, A., *Przem. Chem.* **56**, 599 (1977).
[8] Manes, M., Hofer, L. J. E., *J. Phys. Chem.* **73**, 584 (1969).
[9] Wohleber, D. A., Manes, M., *ibid.* **75**, 3720 (1971).

[10] Chiou, C. C. T., Manes, M., *ibid.* **77**, 809 (1973); **78**, 662 (1974).
[11] Schenz, T. W., Manes, M., *ibid.* **79**, 604 (1975).
[12] Rosene, M. R., Manes, M., *ibid.* **80**, 953 (1976); **81**, 1646, 1651 (1977).
[13] Rosene, M. R., Özcan, M., Manes, M., *ibid.* **80**, 2586 (1976).
[14] Garbacz, J. K., Biniak, S., Świątkowski, A., *Ads. Sci. Tech.* **3**, 61 (1986).
[15] Garbacz, J. K., Cysewski, P., Biniak, A. et al., *ibid.* **3**, 253 (1986).
[16] Deryło-Marczewska, A., Jaroniec, M., *Surface Colloid Sci.* **14**, 301 (1987).
[17] Al-Bahrani, K. S., Martin, R. J., *Water. Res.* **10**, 731 (1976).
[18] Rovinskaya, T. M., *Koll. Zh.* **24**, 215 (1962).
[19] Koganovskii, A. M., Rovinskaya, T. M., *Koll. Zh.* **25**, 447 (1963).
[20] Akhmadeev, V. Ya., Ipatova, E. L., Shevchuk, I. A., *Khim. Tekhnol. Vody* **3**, 405 (1981).
[21] El-Dib, M. A., Badawy, M. J., *Water Res.* **13**, 255 (1979).
[22] Abe, I., Hayashi, K., Kitagawa, M. et al., *Bull. Chem. Soc. Japan* **53**, 1199 (1980).
[23] Boehm, H. P., in: *Advances in Catalysis and Related Subjects* **16**, 179 (1966).
[24] Mattson, J. S., Mark, H. B., *Activated Carbon, Surface Chemistry and Adsorption from Solution,* M. Dekker, New York 1971.
[25] Puri, B. R., in: *Chemistry and Physics of Carbon,* vol. 6, Walker, P. L. Jr. (ed.), M. Dekker, New York 1970.
[26] Puri, B. R., Sharma, S. K., Dosanjh, I. S. et al., *J. Indian Chem. Soc.* **53**, 486 (1976).
[27] Puri, B. R., in: *Advances in Chemistry Series,* No. 202, Treatment of Water by Granular Activated Carbon, McGuire, M. J., Suffet, I. H. (Eds.), American Chemical Society, Washington D. C., 77–93 (1983).
[28] Glushchenko, V. Yu., Levagina, T. G., Pershko, A. A., *Koll. Zh.* **37**, 134 (1975).
[29] Oda, H., Yokokawa, Ch., *Carbon* **21**, 485 (1983).
[30] Oda, H., Kishida, M., Yokokawa, Ch., *ibid.* **19**, 243 (1981).
[31] Coughlin, R. W., Ezra, F. S., *Environ. Sci. Technol.* **2**, 291 (1968).
[32] Ueda, H., Nambu, N., Nagai, T., *Chem. Pharm. Bull.* **28**, 3426 (1980).
[33] Peschel, G., Belouschek, P., Kress, B. et al., *Progr. Coll. Polymer Sci.* **65**, 83 (1978).
[34] Kozlov, S. G., Glushchenko, V. Yu., *Koll. Zh.* **38**, 577 (1976).
[35] Jayson G. G., Lawless, T. A., Fairhurst, D., *J. Colloid Interface Sci.* **86**, 397 (1982).
[36] Koopal, L. K., *Z. Wasser Abwasser Forsch.* **16**, 91 (1983).
[37] Muller, G., Radke, C. J., Prausnitz, J. M., *J. Phys. Chem.* **84**, 369 (1980).
[38] Bauer, U., *Vom Wasser* **39**, 161 (1972).
[39] Bernardin, F. E., Jr., *Chem. Eng.* **83**, 77 (1976).
[40] *Activated Carbon Adsorption of Organics from the Aqueous Phase,* vol. 1, Suffet, I. H., McGuire, M. J. (Eds.), Ann Arbor Science Publishers, Ann Arbor, Mich. 1980.
[41] Fukuchi, K., Hamaoka, H., Arai, Y., *Mem. Fac. Eng. Kyushu Univ.* **40**, 107 (1980).
[42] Fukuchi, K., Arai, Y., *Environ. Conserv. Eng.* **9**, 625 (1980).
[43] Jossens, L., Prausnitz, J. M., Fritz, W. et al., *Chem. Eng. Sci.* **33**, 1097 (1978).
[44] Brown, C. E., Everett, D. H., in: *Colloid Science,* vol. 2, Everett, D. H. (Ed.), The Chemical Society, London 1975.
[45] John, P. T., Ghori, T. A. K., Nagpal, K. C., *Indian J. Technol.* **18**, 261 (1980).
[46] Korta, A., Studium sorpcji substancji polarnych i niepolarnych z par i roztworów wodnych na sorbentach węglowych (Study of sorption of polar and non-polar substances from the vapour phase and aqueous solutions on carbon sorbents), Academy of Mining and Metallurgy, Cracow 1967, Fasc. 13, *Górnictwo* (Mining), No. 189.
[47] Koganovskii, A. M., Levchenko, T. M., *Zh. Fiz. Khim.* **46**, 1789 (1972).

[48] Stadnik, A. M., Él'tekov, Yu. A., *Zh. Prikl. Chim.* **48**, 186 (1975).
[49] Mioduska, M., Pietrzyk, S., Świątkowski, A. et al., *Biul. WAT* **28**, No 9, 109 (1979).
[50] Levchenko, T. M., Kakaulina, T. N., Koganovskii, A. M., *Zh. Fiz. Khim.* **52**, 664 (1978).
[51] Koganovskii, A. M., Salkova, A. A., *Ukr. Khim. Zh.* **38**, 885 (1972)
[52] Stadnik, A. M., Él'tekov, Yu. A., *Zh. Fiz. Khim.* **52**, 2100 (1978).
[53] Stadnik, A. M., Él'tekov, Yu. A., *ibid.* **52**, 1795 (1978).
[54] Stadnik, A. M., Él'tekov, Yu. A., *ibid.* **49**, 2381 (1975).
[55] Stadnik, A. M., Él'tekov, Yu. A., *ibid.* **49**, 1822 (1975).
[56] Stadnik, A. M., Él'tekov, Yu. A., *ibid.* **51**, 2697 (1977).
[57] Jaroniec, M., Choma, J., *Ochrona Środowiska*, **32–33**, 33 (1987).
[58] Jaroniec, M., Madey, R., Choma, J. et al., *J. Colloid Interface Sci.* **125**, 561 (1988).
[59] Koganovskii, A. M., Levchenko, T. M., Kirichenko, V., *Adsorptsiya rastvorennykh veshchestv* (Adsorption of dissolved substances), Naukova Dumka, Kiev 1977.
[60] Ościk, J., *Adsorption*, PWN-Ellis Horwood, Warsaw-Chichester 1982.
[61] Świątkowski, A., Mioduska, M., *Wiadomości Chem.* **35**, 257 (1981).
[62] Jaroniec, M., *Adv. Colloid Interface Sci.* **18**, 149 (1983).

CHAPTER 8

Techniques for Testing the Porous Structure of Active Carbons

8.1 ADSORPTION METHODS

8.1.1 General Remarks

The practical utility of active carbons as adsorbents, catalyst supports or electrode materials is mainly determined by the degree of development of their internal surface. This relates equally to the specific surface area and to the nature of the capillary structure (e.g. the distribution of the pore volume as regards the effective radii). Currently many different research methods are used to obtain information about the configuration of the surface of solids. Among the most important are adsorption and densitometric methods, mercury porosimetry, X-ray radiography and electron microscopy. The adsorption methods play here a specific role in view of the wide spectrum of information they provide, the possibilities of using various adsorbents, different testing procedures and measuring conditions, and the susceptibility of the apparatus to partial or complete automation. The greatest amount of information about the capillary structure of porous adsorbents is obtained from an analysis of the adsorption and desorption isotherms of vapours determined under static conditions. In such tests the following adsorbates are most frequently used: gases such as N_2, Ar, Kr, CO_2 and vapours of low-boiling aliphatic hydrocarbons (both straight-chain and cyclic), alcohols, benzene, carbon tetrachloride, water, etc. One should note that as regards polar adsorbates not only the capillary structure of the adsorbate but also the chemical constitution of its surface (e.g. the presence of oxygen-containing surface groups in the case of adsorption of water vapour) affect the adsorption. Therefore for characterization of the capillary structure such non-polar adsorbates as N_2, Ar or benzene are most commonly used. Today many different kinds of adsorption apparatus operating on various principles are used for determining the adsorption and desorption isotherms. The determination of adsorption at equilibrium and at pressures corresponding to that equilibrium is our concern here. This quantity is found either by direct measurement of the variation of the quantity of adsorbate or from its variation in the space over the adsorbent. Almost every fundamental type of apparatus may have versions differing as regards details of design or level of automation of the operations necessary to determine the adsorption and desorption isotherms. Many details concerning the test procedures relevant to adsorption from the gas phase as well as the descriptions of the design and operation of the apparatus have been given by Gregg and Sing [1].

To illustrate several different methods of determining the adsorption isotherms from the gas phase involving measuring techniques based on different principles, we shall consider in this chapter the McBain and Bakr sorption balances, liquid microburettes and the sorption monostat. Thanks to their relatively simple maintenance and considerable sensitivity these devices find application both in research and industrial laboratories.

8.1.2 Sorption Spiral Balances

Sorption balances with quartz spirals, designed some 60 years ago by McBain and Bakr [2], continue to be used for determining adsorption isotherms of such vapours as benzene, aliphatic hydrocarbons, alcohols or carbon tetrachloride. Their main advantages, apart from simplicity are (i) the possibility they provide, of conducting in parallel, measurements on many samples of adsorbents, and (ii) that during measurements, only the space in which the adsorbent is placed requires thermostating. However, since only small portions of the adsorbent may be tested, sufficiently high accuracy is achieved only for adsorbents whose sorption is not too small. The design of the sorption spiral balance is illustrated in Fig. 8.1 where one version is shown schematically. Apart from the principal elements (specified in the figure caption) the setup usually includes an impeller vacuum pump, a diffusion pump, cold finger initial vacuum tank (these devices are not shown in the figure as their function is the degassing of the adsorbate), generation of a vacuum in the measuring system, and removal of adsorbate vapours when determining the desorption branch of the isotherm. Similar devices also constitute parts of sorption apparatus of different types where they fulfil the same functions. In typical laboratory sorption balances,

Sec. 8.1] **Adsorption methods** 195

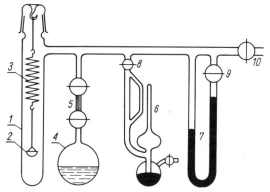

Fig. 8.1 Scheme of the McBain and Bakr vacuum sorption balance with quartz spiral: 1 — glass tube, 2 — dish with adsorbent, 3 — quartz spiral, 4 — vessel with liquid adsorbate, 5 — sluice for admitting small quantities of adsorbate vapours, 6 — McLeod gauge, 7 — mercury manometer, 8, 9, 10 — valves.

measuring systems are used consisting of at least 2–3 glass tubes with quartz spirals connected to a common vacuum line. This part of the apparatus is maintained at constant temperature by placing it in a thermostated chamber made of transparent material.

The following procedure is used to determine the adsorption and desorption isotherms by means of sorption balances. The measurement is preceded by degassing of the liquid adsorbate under vacuum. For this purpose the vacuum pump is switched on, valves 5 and 10 being kept open. The next operation is calibration of the quartz spirals by measuring with a cathetometer their elongation under various known loads. The position of the spiral is measured with respect to a constant point, e.g. the end of a bar fixed to the coil of the spiral. Next a weighed portion of adsorbent is placed in a basket hanging at the end of the spiral and is subject to degassing under a pressure of the order of 10^{-3} Pa and at a temperature of 200–400°C. During this operation valves 5 are closed, and the glass tubes with the spirals slipped into ceramic tube electric furnaces. After degassing is concluded (a constant pressure of $\leqslant 1.3 \times 10^{-3}$ Pa is established) valve 10 is closed and the furnaces are replaced by a thermostated chamber keeping the adsorption space of the apparatus at the required constant temperature. Now the positions of the ends of all the spirals are determined using a cathetometer and the measurement itself is started. The quantity of adsorbate vapours is measured in the sluice by successive opening and closing of its two valves

5. After the adsorption equilibrium is established, the elongation of the spiral and the corresponding equilibrium pressure of the adsorbate vapours are read. For small pressures corresponding to the initial part of the isotherm, use is made of the McLeod gauge.

The attainment of adsorption equilibrium is evident from a given elongation of the spiral and the establishment of equilibrium pressure. The time taken to achieve equilibrium may vary both with the type of adsorbent and on its equilibrium pressure. Successive points of the isotherm are determined by introducing known quantities of the adsorbate vapours and reading the elongation of the spiral and pressures after equilibrium is established. In the higher pressure range, a mercury manometer and cathetometer are used, and the portions (ever greater) of adsorbate are measured by opening both valves 5 of the sluice simultaneously. When determination of the entire adsorption isotherm, i.e. the relationship between adsorption (proportional to the elongation of the spiral) and adsorbate equilibrium pressure is complete, the desorption branch of the isotherm is determined. The pressure of the adsorbate at the final point of the adsorption isotherm is usually very close to the adsorbate saturated vapour pressure at the temperature of measurement. Successive points of the desorption isotherm are determined, starting from this point, by switching on the vacuum pump and opening for a short time valve 10 in order to remove from the measuring part of the apparatus a portion of the adsorbate vapour. After equilibrium is established (i.e. when part of the adsorbate is desorbed) the shortening of the spiral and the equilibrium pressure are measured. It should be noted that the sorption apparatus described is only one possible version. The key elements, e.g. the device for adsorbate dosing, the fixing of the spiral, type of valves, manometers, etc. may differ considerably in design, although the overall design and principle of operation remain the same. If necessary, e.g. as regards the adsorbate, the valves and grindings (in which vacuum grease is used) and also the mercury manometers may be completely eliminated. In this case, however, the apparatus must be sealed after the necessary vacuum is generated. Such a procedure is used in examining the adsorption of halogens on silica gel and active carbon [3].

8.1.3 Adsorption Liquid Microburettes

The method involving the use of liquid microburettes in adsorption measurements has been known for over sixty years [4] but its application became widespread only after *ca.* 1950 (chiefly in Soviet laboratories) [5, 6]. Microburettes are particularly suited for determination of the adsorption isotherms of substances whose saturated vapour pressure is not too small at temperatures not far from room temperature, e.g. water, methanol and ethanol, cyclohexane and benzene. In this method adsorption is determined from the lowering of the liquid meniscus (of the adsorbate) in a calibrated microburette, correction being made for the dead volume of the apparatus. The accuracy of this procedure is in many cases greater than that of the gravimetric method because of the possibility of using fairly large amounts of the adsorbate.

As an example of a microburette adsorption apparatus we shall consider the setup designed in the early sixties at the Academy of Mining and Metallurgy in Cracow [7] which is a modification of earlier versions. The setup considered is notable for its simple design and for having novel constructional details facilitating particularly the measurement of desorption equilibria. Two versions of the apparatus are available. The first is suitable for determination of adsorption isotherms of adsorbates whose vapours do not react with vacuum grease (e.g. water, methanol), while the second is universal and may be used for determining both sorption and desorption isotherms of vapours of both polar and non-polar molecules. The two types of apparatus are shown schematically in Fig. 8.2. Since they have many features in common we shall discuss their design together. The main item is the measuring capillary C which is 20 cm long and of *ca.* 1.5 mm internal diameter. Prior to mounting the capillary in the setup, the uniformity of its diameter is tested and it is accurately calibrated with mercury, so the volume of a 1 mm section of the capillary can be determined. Marks are etched on the capillary at 50 mm intervals which act as reference points for the measuring the position of the liquid meniscus with the cathetometer. The measuring capillary C is connected at the bottom to a microcapillary mC about 0.1 mm in diameter. At the top, capillary C is connected to a bulb B of 2 mm diameter. At its upper section (below the bulb), the capillary is wrapped in a disposable, moistened cotton-wool collar T. The other features in common to both types of apparatus are the ampoule A, in which the adsorbent is placed, the mercury manometer M (in the case of water vapour adsorption silicone oil can also be used as the manometer liquid), and vacuum stopcock S_1 connecting the apparatus via the right arm of the manometer to the vacuum line. The differences in design are related to the different types of adsorbate used, and refer to such details as the attachment of the ampoule A and of stopcocks S_2 and S_3. As regards the former, if the adsorbate used is a liquid that does interact

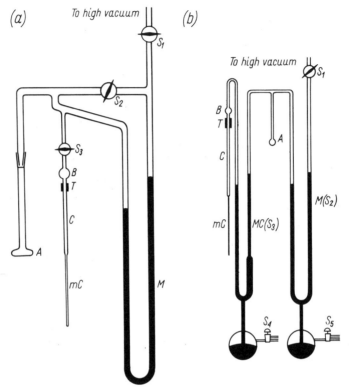

Fig. 8.2 Microburettes for determining adsorption: (a) with conventional vacuum stopcocks in the adsorption section, (b) with mercury cut-offs (universal version) (after Lasoń and Żyła [7]).

with vacuum grease, then ampoule A is connected by means of ground glass joints; otherwise the ground joint must be eliminated by connecting the filled ampoule directly by fusion. The second difference also relates to sensitivity of the adsorbate towards vacuum grease, and consists of replacing the conventional (ground) stopcocks by mercury cut-offs. The mercury cut-off S_3 like the conventional vacuum valve S_3 cuts off the measuring capillary from the adsorption section of the setup. The mercury cut-off is opened on lowering the level of mercury by connecting stopcock S_4 to the vacuum. To facilitate the transfer of adsorbate vapour from capillary C to the measuring section of the apparatus, the right arm of the mercury cut-off is widened along a section of about 15 cm. The role of the second vacuum stopcock S_2 is taken over in the universal design of the setup by manometer M in which it is possible to remove mercury from both branches by opening stopcock S_5

and switching on the vacuum pump operating in the same way as in the mercury valve.

For adsorption tests in practice, usually sets of both types of microburette consisting of 4, 8 or 10 setups (each consisting of a capillary, ampoule, manometer and stopcocks) are used. Such a set, attached to a common high vacuum line, is enclosed in a transparent plastic housing and thermostated with air with an accuracy up to $\pm 0.1°C$.

The determination of the adsorption and desorption isotherms using this apparatus involves the following operations.

Filling of the measuring capillary C with the adsorbate and its preparation for taking measurements. The open end of the microcapillary is immersed in a vessel containing the adsorbate, and by decreasing the pressure slightly, the liquid is introduced into the capillary up to the bulb. After rinsing the capillary several times, it is filled with adsorbate to half its height. Next, after closing stopcock S_3, the microcapillary is sealed above its end. The gases dissolved in the adsorbate and the adsorbate vapours are removed by applying a vacuum, preferably from an independent vacuum equipment, until the liquid column in the measuring capillary and in the microcapillary, reaches a constant level (stopcock S_3 whould be opened with caution when stopcocks S_1 and S_2 are open). Next, stopcocks S_1 and S_2 are closed, and the moistened cotton-wool collar is wrapped around the measuring capillary just beneath the liquid meniscus. The slowly evaporating water slightly reduces the temperature of the capillary in the vicinity of the meniscus, preventing distillation and condensation of the adsorbate beyond the measuring capillary. A single filling of the microburette usually suffices for determining 2–3 adsorption isotherms, depending on the quantity and properties of the adsorbent.

Before starting the measurements proper, a curve should be constructed giving the variation of the lowering of the meniscus in the capillary with the pressure of the adsorbate vapours in the adsorption volume. This correction, which relates to the dead volume of the apparatus, is applied when determining the isotherm.

Preparation of the adsorbent. All the operations associated with placing the adsorbate in the ampoule are carried out after closing stopcocks S_1 and S_3, and opening stopcock S_2. The degassing of the adsorbent sample (usually combined with heating, when a small temperature-controlled electric heater is placed over the ampoule) is conducted with stopcocks S_1 and S_2 open and stopcock S_3 closed until the pressure is lowered to 1.3×10^{-3} Pa. If heating of the adsorbent is undesirable, rinsing with helium is applied [7].

Adsorption measurements. After the preliminary operations are concluded, stopcocks S_1 and S_2 are closed, and the thermostating system is switched on. When the temperature becomes constant, the distance of the liquid meniscus from the mark on the capillary is measured with an accuracy up to 0.02 mm by means of the cathetometer. The opening of stopcock S_3 causes the distillation of some liquid from the measuring capillary to the adsorption volume. The quantity of the vapour adsorbed is determined from the lowering of the liquid meniscus in the capillary, following correction for the dead volume of the apparatus. The pressure of the adsorbate vapour at adsorption equilibrium, established after stopcock S_3 is closed, is read from the manometer M with an accuracy of 0.02 mm. The quantity of adsorbate adsorbed by 1 g of the adsorbent is calculated from the equation:

$$a = \frac{[\Delta h - \Delta h'(p)] sd}{mM} + \frac{pV'}{mRT} \qquad (8.1)$$

where a is the quantity of adsorbate adsorbed in mol g^{-1} of adsorbent, Δh is the difference in levels of the adsorbate meniscus in the microburette, $\Delta h'(p)$ is the correction for the dead space, s is the capillary constant (i.e. the volume of the length of 1 mm of the capillary), d is the density of the adsorbate, M is the molecular mass of the adsorbate, m is the mass of the adsorbent sample, pV'/RT is the correction for the volume of adsorbent, V is the volume of the adsorbent sample, and p is the pressure of the adsorbate at adsorption equilibrium.

To determine the desorption isotherm we proceed as follows. After the last point of the adsorption isotherm has been found, stopcock S_3 is opened and the bulb B is wrapped in a cotton-wool collar moistened with a volatile liquid (e.g. water, alcohol, acetone). The desorbing vapours condense on the walls of the bulb. After closing stopcock S_3, the condensate is made to descend with the help of the collar to the measuring capillary. The quantity of the adsorbate that has desorbed is calculated from the difference in the levels of the liquid meniscus in the capillary before and after desorption, and the pressure at equilibrium is read from manometer M.

8.1.4 Adsorption Manostat

A very useful tool for adsorption tests is a device whose operation is based on the principle of the manostat. We shall discuss here a semi-automatic sorption manostat designed in the early seventies at the Academy of Mining and Metallurgy in Cracow [8]. This apparatus enables the determination of the adsorption and desorption isotherms of various gases and vapours at specified temperatures, and especially those of nitrogen, argon, and oxygen at 77 K and of carbon dioxide at 195 K, as well as that of the adsorption kinetics of these adsorbates over the entire range of relative pressures. The principle of operation is based on the determination of the points of the isotherm at predetermined pressures kept constant during one measurement. The pressure drop due to adsorption is compensated by lifting the mercury column in the measuring burette, whereas the pressure rise due to desorption is compensated by lowering the mercury column in the burette. Both of these processes are automatic.

The apparatus design is presented schematically in Fig. 8.3. The essential, original features in this layout are the automatic mercury cut-off, which enables admission of mercury to the gas burette during adsorption (as well as its removal from the burette during desorption), and the contact probe. A part of the apparatus (enclosed by the dashed line in Fig. 8.3) is situated in a transparent plastic housing and air thermostated.

The procedure applied when determining the adsorption isotherm by means of the semi-automatic adsorption manostat is as follows.

Prior to taking measurements, the test sample of adsorbent situated in an ampoule is degassed until the pressure reaches a constant value of $ca.$ 10^{-3} Pa. The pressure is measured by means of a McLeod gauge. In the degassing process the stopcocks 16, 17, 14, 12, 19 and 15 are open, and the temperature of the sample, depending on its properties, is maintained by placing a small electric heater over the ampoule. If heating is inadvisable, then good results are obtained by rinsing with helium.

After the pressure has been lowered in the ampoule to the required level ($\leqslant 1.3 \times 10^{-3}$ Pa), the stopcocks 16, 14 and 12 are closed and by opening the stopcocks 22 and 13, gaseous adsorbate is introduced into the apparatus until the required pressure, measured on manometer 5, is achieved. The pressure may be established more accurately by making use of that part of apparatus zoned within the stopcocks 11, 12 and 14. Mercury in the gas burette is lowered to the zero level before the measurement is started. The ampoule with the adsorbent is immersed in a Dewar filled with liquid nitrogen. It is important that the contact sensor 4, controlling the automatic mercury cut-off via the electronic

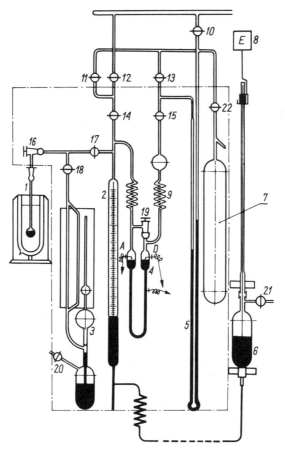

Fig. 8.3 Semi-automatic manostat for determining adsorption: 1 — ampoule with adsorbent, 2 — calibrated gas burette, 3 — McLeod gauge, 4 — contact sensor controlling mercury cut-off, 5 — mercury manometer, 6 — automatic mercury cut-off, 7 — container with adsorbate, 8 — electromagnet, 9 — glass spiral, 10–22 — vacuum stopcocks.

relay, be regulated prior to the measurement. The regulation is performed by opening stopcock 19 and shifting (a process enabled by the glass spirals 9) the lower part of the probe along a beam until the mercury meniscus in the right- or left-hand bulb (depending on whether adsorption or desorption is being carried out) is brought into contact with the upper end of the contact wire.

When the mercury level in the sensor is etablished, stopcock 19 is closed, which produces a separation of the measuring part of the setup (ampoule, burette) from the rest, the electric circuit (relay and magnet) is

Sec. 8.1] **Adsorption methods** 203

switched on which produces an automatic setting of the device. In the sensor a minimum spacing of the contact takes place, and the mercury in the measuring burette is established at a level taken as initial. The measurement proper is started by opening stopcock 16 (in the case of kinetic tests a stop-watch is started simultaneously). This causes an insignificant pressure drop in the section due to filling of the ampoule volume by the adsorbate, and simultaneous adsorption. In the sensor the wire is brought into contact with the mercury meniscus which triggers (via the electronic relay) the electromagnet into operation. As a result, the mercury cut-off is opened and a definite amount of mercury is introduced into the measuring burette. The sensor therefore returns to its initial position. The level of mercury in the burette is then raised stepwise. The changes of the mercury level as a function of time reflect the adsorption kinetics.

The quantity of gas adsorbed at the specified pressure is found from the difference between the mercury levels in the burette before and after the measurement, expressed in terms of volume, and corrected for by the dead space of the ampoule (the difference of temperatures between the burette and ampoule being accounted for). After the adsorption equilibrium is established, stopcock 16 is closed, the magnet switched off, and after opening stopcock 19, a new pressure is established. The cycle of operations is then repeated.

If the quantity of gas adsorbed is very large at a particular point of the adsorption isotherm, it is sometimes necessary to refill the burette with the adsorbate. For this purpose stopcock 16 is closed and the sensor stopcock 19 kept closed. By opening stopcock 21 with the help of the impeller pump, the pressure is lowered over the mercury in the bulb of the mercury cut-off. Hence the opening of this cut-off produces a flow of mercury from the burette to the cut-off container. In place of the mercury removed, a suitable quantity of adsorbate is introduced into the burette (through stopcocks 11 and 14). By subsequently switching on the electric circuit, the pressure is equalized automatically at the level that existed in the apparatus before the procedure was started.

The procedure described relates to the determination of adsorption (points on the adsorption isotherm) only when the equilibrium pressures of the adsorbate are not too low. When points need to be determined on the isotherm for small pressures (indispensable especially when testing active carbons) the procedure is somewhat modified. Use is then made of the McLeod gauge included in the measuring part of the apparatus. After accurate degassing of the adsorbent and having closed stopcocks

16 and 17 and keeping closed stopcock 18, we fill the apparatus with adsorbate at the required pressure (usually $4 \times 10^{-3} - 5 \times 10^{-3}$ Pa) and make preparations for determining the points of the adsorption isotherm as described in the standard procedure given above. The adsorbate is introduced through stopcock 17 into the space between stopcocks 16, 17 and 18, and its quantity is determined from the change in the level of mercury in the gas burette. After stopcock 17 is closed and stopcocks 16 and 18 are opened, the gas expands to the volume of the McLeod gauge and ampoule, where it is adsorbed on the adsorbent. When adsorption equilibrium is established, stopcock 16 is closed and the pressure corresponding to that equilibrium is measured with the McLeod gauge. Knowing the pressure in, and volume of, this part of apparatus, we can calculate the quantity of gas remaining after adsorption is complete. The quantity of gas adsorbed (for the given point of the adsorption isotherm) is the difference between the quantity of gas initially introduced through stopcock 17 and that left in this part of the apparatus after adsorption equilibrium has been established.

The adsorption isotherm having been determined (usually under pressures corresponding to a relative pressure of 0.98), one often determines the desorption isotherm. The appropriate procedure is as follows. The level of mercury in the gas burette should be maximal. Stopcock 16 being closed and 19 open, a certain quantity of gas is removed from the apparatus through stopcocks 14 and 12 to achieve the predetermined pressure. To increase the accuracy of measurement the switch of the sensor is moved from position A (adsorption) to position D (desorption), upon which the sensor is regulated according to the scheme given above. Next, after closing stopcock 19 when a preliminary vacuum is generated in the container of the mercury cut-off, and stopcock 21, we switch on the electric circuit which produces automatic setting of the apparatus. On opening stopcock 16 the desorbing gas evolves, producing a minimal increase of pressure in the measuring part of the apparatus. Then the sensor switch D is brought into contact with the mercury, which produces in turn the opening of stopcock 6 and evacuation of a certain quantity of mercury from the measuring burette and re-establishment of pressure at the initial level. The measurements of the variation of the mercury level in the burette as a function of time give us a picture of the desorption kinetics. In concluding this description of the sorption manostat, we should note that these devices are usually used in sets of two or four connected to a common vacuum line, adsorbate container and mercury manometer.

8.1.5 Automation of Adsorption Testing Apparatus

The variety of applications of carbon adsorbents is ever-expanding and the need for improved knowledge of their properties brings about constant developments in the design of apparatus for adsorption testing. This relates especially to apparatus used for determining adsorption-desorption isotherms of gases and vapours, both by gravimetric and volumetric methods. Such improvements have the primary aim of automating the necessary operations. As a result, the number of operations is reduced and the manipulations become less labour-intensive, the measuring errors are smaller and more objective, the measuring cycle may be programmed, and the manometric liquids can be eliminated.

In the case of gravimetric equipment, automation of the apparatus requires incorporation of a device recording the responses of the sorption balance, an automatic system for admitting the gaseous adsorbate, a system for control and recording of the adsorbate pressure, and a programming centre. The commonest designs are usually partially automated. Most often they have an automatic balance, for instance the Gregg sorption balance [9] in which the deflection of the beam is compensated by the increase in the force of mutual attraction between the solenoid and the constant magnet attached to the balance, due to the increase of the current flowing through the coil of the electromagnet. This design, known for over 30 years, continues to be the basis for building electronically controlled, fully automatic adsorption balances. The most frequently-used method for automating weighing is in this case an optoelectronic system of compensating and signalling the balance deflection. The optical part of the system includes in most adsorption apparatus, light sources, mirrors and photoelectric detectors. In many designs induction sensors of the balance deflection are used. The balance beams are often suspended on thin tungsten tape or thread.

Another development in automatically recording sorption balances is the modification of conventional quartz or metal spring balances to incorporate electronic devices recording elongation. As in automatic beam balances, optical or induction recording systems are customary. The designs of sorption balances based on quartz spirals are particularly suitable for handling the adsorption of chemically aggressive vapours and gases as they have no components made from readily-corroded materials. Another advantage of spring balances is that almost all components of the measuring and recording system are situated outside the vacuum system which facilitates the achievement of low pressures when preparing the adsorbent for the tests (heating combined with desorption). One drawback of set-ups with spring balances is that the variation of the spiral elongation with load is not always linear, and that the maximum permissible weights of the adsorbent samples are therefore relatively small. Detailed information on vacuum adsorption balances can be found in the literature [10, 11].

Attempts have also been made to automate volumetric apparatus. These refer especially to automatic admission of the gaseous adsorbate to the ampoule containing the adsorbent, recording of the equilibrium pressure of the vapours or gases, as well as the pressure of the dosed medium, and the programmed control of the measuring cycle. These approaches apply both to the apparatus operating as a manostat and to that in which the pressure of the adsorbate vapours is different at the moment of exposure to the adsorbent and after equilibrium is established, and the volume of the adsorbent is the constant parameter. Currently there are many adsorption set-ups of different design which are largely automated. These refer not only to laboratory-designed set-ups but also to those commercially-available (e.g. the Sorptomatic unit manufactured by Carlo Erba). More information on semi-automatic and automatic volumetric adsorption testing apparatus can be found in the literature [10].

In recent years automation of adsorption measuring apparatus has been combined with its computerization. Incorporation of a computer enables on the one hand optimal programming of the measuring cycle, and on the other, direct processing of the experimental data. In effect the time necessary for carrying out measurements has been drastically shortened and information about the porous-structure characteristics of the adsorbent can be obtained directly.

One should note that all the devices considered above for measuring adsorption operate on the basis of static methods of determining adsorption.

8.1.6 Methods of Determining Adsorption under Dynamic Conditions

The dynamic methods of determining adsorption from the gas phase on solid adsorbents usually require less complicated (vacuum-free) apparatus as compared with static methods. Under dynamic conditions adsorption is determined by passing at a constant rate and temperature, a mixture of carrier gas and adsorbate (e.g. vapours of an organic substance) through a column containing a known quantity of adsorbent. The concentration of the adsorbate in the gas stream is usually varied by mixing, in varying proportions, the pure carrier gas (e.g. air) with the same carrier gas saturated with vapours of the adsorbate. Successive points of the isotherm are found by saturating the adsorbent samples with the adsorbate to constant mass. This method is most suitable for testing granulated or granular, but not powdered, adsorbents. More details on the design of such apparatus can be found in the literature [12].

Among the dynamic methods we should note especially adsorption gas chromatography [13]. Adsorption tests by this method can be conducted over a wide range of pressures and temperatures. Analysis of the chromatograms not only allows us to determine the adsorption isotherm but also to obtain data on the kinetics of the adsorption and desorption processes. A certain limitation of the chromatographic methods arises from interference by diffusion processes. Good agreement is obtained between adsorption isotherms determined by chromatographic and static methods when the pore radii are not too small.

8.2 TOTAL PORE VOLUME. REAL AND APPARENT DENSITY

The total pore volume V_Σ, is an important quantity characterizing adsorbents (including active carbons). It is the sum of the volumes of the three types of pore occurring in adsorbents:

$$V_\Sigma = V_{mi} + V_{me} + V_{ma} \qquad (8.2)$$

where V_{mi} is the volume of the micropores, V_{me} is that of the mesopores and V_{ma} that of macropores. The volumes of the micro- and mesopores are determined from adsorption measurements (i.e. gas or vapour adsorption and desorption isotherms). However, the determination of the volume V_{ma} requires knowledge of the total pore volume,

$$V_{ma} = V_\Sigma - (V_{mi} + V_{me}). \qquad (8.3)$$

The total pore volume is found on the basis of the apparent and real adsorbent density. Since for determining these densities mercury and helium are usually used, they are referred to as mercury d_{Hg} and helium d_{He} densities. The total pore volume is equal to the difference of the reciprocals of these quantities:

$$V_\Sigma = \frac{1}{d_{Hg}} - \frac{1}{d_{He}}, \qquad (8.4)$$

208   **Testing the porous structure**   [Ch. 8]

The mercury method of determining apparent density is based on the fact that under atmospheric pressure mercury does not penetrate into pores whose diameter is smaller than about 1.5×10^4 nm, the shape of the adsorbent particles having little effect on the results of measurements. The apparatus used for determination of the apparent (mercury) density is simple and easy to use [14]. The principal components of the apparatus (Fig. 8.4) are the ampoule A with the adsorbent, the mercury container C, stopcocks S_1 and S_2, and connecting tubes. Ampoule A is of a special boot-like shape, preventing the adsorbent from emerging above the mercury surface, and is connected with the apparatus by means of a ground joint G.

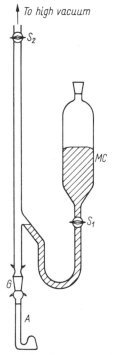

Fig. 8.4 Apparatus for determining the apparent density of porous materials by the mercury method (after Czapliński [14]).

The measuring procedure is as follows. An accurately weighed portion of the preliminarily-dried adsorbent is placed in ampoule A which is connected with the measuring apparatus by the ground glass joint G. The apparatus is then connected to high vacuum by closing stopcock S_1 and opening stopcock S_2. When a vacuum of the order of

10^{-2} Pa is established, mercury is introduced slowly into the ampoule from container C through stopcock S_1 until its meniscus is located somewhat above the mark etched on the neck of the ampoule; then stopcock S_2 is closed and the ampoule disconnected from the apparatus. The ampoule with mercury and the adsorbent is placed in a water thermostat usually heated to 20° C. The excess of mercury above the mark is removed from the ampoule by using a glass tube ending with a capillary, the upper end, provided with a bulb for collecting the mercury, being connected to a vacuum. Next the ampoule is weighed. The weight of the ampoule with mercury only (without the adsorbent) is determined in an identical manner. The apparent density is calculated from the formula:

$$d_{Hg} = \frac{13.546 m}{a - b + m} \qquad (8.5)$$

where 13.546 g cm^{-3} is the density of mercury at 20° C, m is the mass of the adsorbent sample, a is the mass of the ampoule with mercury only (without the sample), b is the weight of the ampoule with mercury and the adsorbent sample. The necessity of using adsorbents with particle diameters not smaller than 0.7 mm somewhat limits the method.

For determining the real (helium) density of adsorbents, the assembly designed at the Academy of Mining and Metallurgy in Cracow proves very useful (Fig. 8.5) [15]. The principal parts of this apparatus are as follows: the ampoule A in which the test adsorbent is placed, the gas burette B with two bulbs of known volume (the burette is filled with mercury with the help of the mercury cut-off MC_1, stopcock S_1 and more accurately stopcock S_2), the measuring manometer MM (with a glass hook fused to the inside wall of the left arm enabling accurate setting of the same mercury level with the help of the mercury cut-off MC_2 in conjunction with stopcock S_3), mercury manometer M, helium container HC, helium dosing device HD consisting of the mercury cut-off MC_3 and stopcock S_6 and for more precise dosing, stopcock S_5, and finally stopcock S_4 connecting the apparatus to the high vacuum line and simultaneously to a McLeod gauge.

The measurements are taken according to the following procedure. An accurately weighed portion of the adsorbent is placed in the ampoule A and degassed with heating with only the mercury cut-off MC_2 and stopcock S_4 open. In some cases flushing with helium is a preliminary operation. After the required vacuum (pressure $< 10^{-2}$ Pa) is achieved, stopcock S_4 is closed and the measurement is started. The level of

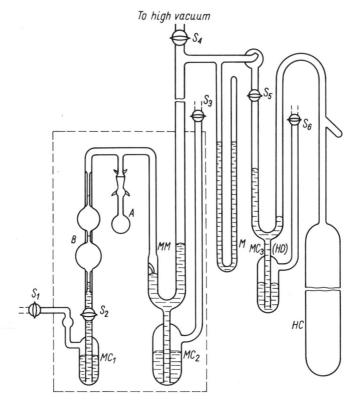

Fig. 8.5 Apparatus for determining the real density of porous materials by the helium method (after Czapliński and Lasoń [15]).

mercury in burette B is set at the etched mark below the lower bulb. With the help of mercury cut-off MC_3 (helium dosing device) helium is admitted to the measuring system through stopcock S_5 until the required pressure is established as read from manometer M. Next the helium-dosing system is cut off and the level of mercury is raised in the measuring manometer MM until it reaches the end of the hook in the left arm. The pressure in the measuring part of the apparatus is read with a cathetometer (from the difference in mercury levels in manometers MM and M).

Subsequent operations consist of filling the lower bulb to the mark with mercury with the help of mercury cut-off MC_1 and with stopcocks S_1 and S_2, followed by re-establishing the level in the left arm of the measuring manometer MM (at the level of the end of the hook) and determining (by means of a cathetometer) the levels of mercury in the right arm of manometer MM and in both arms of manometer M (which allows us to determine the pressure in the measuring system). Similar operations are repeated after filling the upper bulb of the burette with mercury. Thus we obtain information about the changes of helium pressure in the measuring part of the apparatus due to the decrease of its total volume by accurately determined values (i.e. the known volumes of the bulbs). A part of that volume is occupied by the adsorbent sample of accurately known weight. If we repeat these operations without the adsorbent, and apply Boyle's law, $p_1 V_1 = p_2 V_2$ (the measuring system, enclosed in the dashed-line frame, being thermostated), the volume of the adsorbent sample can be accurately determined, and hence its density. Since helium has the smallest atomic radius of the various liquids and gases used in these kinds of test, it can fill even the smallest micropores with diameters as little as 0.25 nm; besides, under the measurement conditions practically no adsorption occurs. One can assume therefore that the densities obtained for the adsorbent are very close to the real values. Small discrepancies are possibily due to the presence in the adsorbent of closed pores inaccessible to helium. In some cases we may have to contend with the inconveniently slow filling of the smallest micropores by diffusion. The density of graphite of 2.26 g cm^{-3} is the limiting value of the real density of substances of (ash-free) organic origin contained in carbonaceous materials. Korta et al. [16] derived and verified a formula for calculating the real (helium) density of the active carbons derived from the ash-free substance contained in carbonized materials of organic origin. This formula has the form:

$$d_{He(org)} = \frac{d_{He}(1-b)}{1 - \frac{d_{He} b}{d_{He(min)}}} \tag{8.6}$$

where d_{He} is the helium density of the carbonizate or active carbon in g cm^{-3}, $d_{He(min)}$ is the helium density of the mineral substance (ash) present in the carbonizate or active carbon in g cm^{-3}, $d_{He(org)}$ is the helium density of the substance of organic origin present in the carbonizate or active carbon in g cm^{-3}, b is the proportion of mineral substance (ash) in the carbonizate or active carbon in g g^{-1} and $(1-b)$ is the proportion of the substance of organic origin in the carbonizate or active carbon in g g^{-1}.

Korta et al. have also shown [17] that the helium density of the substance of organic origin contained in extruded active carbons increases with increasing activation temperature and degree of burn-off.

Additionally, these authors found that for carbon-tar carbonizate in the form of cylindrical granules (of cross-sectional diameter 1.4 mm) and following contact with helium for about 0.5 h (the time usually involved in measurements of helium density) the value of $d_{He(org)}$ is equal to 1.963 g cm^{-3}. In the case of the same carbonizate but ground to a particle size of 10^{-4} cm and kept in contact with helium for 6 days the value found for $d_{He(org)}$ was 2.176 g cm^{-3} which is close to the value for $d_{He(org)}$ found for carbon-tar carbonizate activated at 950° C with a degree of burn-off of about 0.7. This experiment proves that:

(a) the carbonizate granules contain voids virtually inaccessible even to helium,

(b) grinding of the carbonizate granules and long exposure to helium has the same effect as activation, i.e. the voids are made accessible to helium.

The process of converting these voids to pores accessible to helium, which takes place during activation of the granules, leads to a decrease in the volume of their skeleton inaccessible to this gas but without any loss of weight. This in turn produces an increase in the ratio of the weight of the substance of organic origin to its volume still inaccessible to helium, i.e. an increase of the observed helium density $d_{He(org)}$ of this substance.

The authors believe that the increase of $d_{He(org)}$ taking place during activation with increasing burn-off, is also to some extent due to the fact that the more reactive 'amorphous' component, which is less spatially-ordered and of lower density, is the first to burn-off, as a result of which the proportion of the less reactive 'crystalline' component of greater spatial ordering (and greater density) increases in the residual material.

It deserves noting that many other methods are known of determining the density of active carbons and other adsorbents. Apparatus of different designs is also used. More information on this subject may be found in monograph [18]. The methods described above are widely used in Polish laboratories since they display high accuracy, good reproducibility, and simplicity in the design and operation of the apparatus involved.

8.3 MERCURY POROSIMETRY

Adsorption methods based on the phenomenon of capillary condensation find the widest application in tests of the texture of porous solids. However, since their applicability is limited to the range of effective pore radii of 1.5–200 nm, their application to pores with greater effective radii, encounters significant experimental difficulties, as the condensation of vapours then proceeds at a pressure close to saturation vapour pressure and sufficient accuracy of the measurements is then impossible. Adsorption methods can then be supplemented by the method consisting of forcing into the pores a liquid (usually mercury) that fails to wet the adsorbent surface. In this method, known as mercury porosimetry, the volume of mercury is determined which, under a known excess pressure, fills the pores. The forcing of mercury into capillaries of ever-smaller radii requires the application of ever-higher pressures. This relationship is described quantitatively by the equation due to Washburn which may be presented in the following form convenient for the interpretation of results:

$$r = \frac{2\sigma \cos \theta}{p} \qquad (8.7)$$

where r is the effective pore diameter (for capillaries assumed cylindrical) filled with mercury under pressure p, σ is the surface tension at the mercury-adsorbent phase boundary (for carbons a value of 0.480 N m^{-1} is usually assumed) and θ is the wetting angle of the adsorbent by mercury (for carbon adsorbents a value of 2.478 rad is usually assumed). Currently porosimetric measurements are usually made with commercial devices (in many laboratories those marketed by Carlo Erba are used). The results of measurements are generally given in the form of the relationship between the volume of mercury filling the pores (in cm^3 g^{-1}) and the effective radii or the logarithms of the pore radii, or in the form of a plot of $\Delta V/\Delta \log r$ versus $\log r$ giving the differential distribution of the pores according to their effective radii. The range of pore radii for which the above relationships may be determined by mercury porosimetry is approximately 3–7500 nm. As can be seen, we obtain in this way the characteristics of meso- and macropores which supplement data derived from adsorption measurements.

Mention should be made of some limitations of the porosimetric method. The first is associated with the simplifying assumption (on the basis of which equation (8.7) was derived) that the pores are cylindrical. Since this is usually not so, and the shape of the pores differs significantly from cylindrical, the effective radius is considered. The second significant limitation relates to mercury as the pore-filling liquid, and is connected with the lack of numerical data regarding both the surface tension of mercury and the angle of wetting the adsorbent by mercury. The values assumed relate to mercury of great purity. However, in practice mercury is brought into contact with an impure adsorbent surface and becomes contaminated, experiencing therefore a change in its surface tension. Changes in the surface properties also cause changes of the wetting angle. It should be noted that the applied pressure has also some effect on the quantities σ and θ. In view of these limitations, the results of porosimetric tests may carry a systematic error. The last important limitation is the possibility (especially at high pressures) that permanent or elastic deformations of the pore walls may occur. The degree of vulnerability of the pores varies with the type of porous material tested. Characteristics of the meso- and macroporous structure of the adsorbent obtained from mercury porosimetry corresponds (under all these various limitations) to the real capillary structure only if the wider pores, access to which is from the external surface of the granules, pass gradually into ever-narrower capillaries. In the opposite situation, i.e. when narrower pores pass gradually into wider ones, mercury porosimetry yields completely false results.

Concluding, despite the shortcomings of mercury porosimetry, this method finds wide application in testing adsorbents (including active carbons). It provides rapidly information about the pore volume and the distribution of pore volumes in terms of the effective pore radii over a wide range of radii.

8.4 X-RAY SMALL-ANGLE SCATTERING METHOD

The first attempts to apply small-angle scattering of X-rays to the study of the porous structure of active carbons were made in the forties and had a qualitative character. The scattering of X-rays at small angles (i.e. in the range 10^{-4}–10^{-1} rad) occurs in media with electron density inhomogeneities the size of which varies within the range 1–100 nm. In the case of adsorbents (active carbons), the pores inside the solid phase may constitute such scattering centres. Analysis of the curves describing the dependence of the scattering intensity I on the angle φ provides information on the magnitude, and in many cases also on the shape, of the inhomogeneities of the solid. In the fifties the following equation was developed [19] for describing the intensity distribution of X-ray scattering:

$$I_\varphi = I_e N n^2 \exp-\frac{4\pi^2 R \varphi^2}{3\lambda^2} \tag{8.8}$$

where I is the intensity of scattering at angle φ, I_e is the intensity of scattering for one electron, λ is the wave length of the X-ray radiation used, N and n are the number of inhomogeneities in the solid scattering the X-rays and the number of electrons in each of them, respectively and R is the inertia radius, characterizing the mean numerical dimensions of the inhomogeneity.

The determination of the inertia radii for a range of saccharose-based active carbon samples with different degrees of activation provides an example of the application of the above equation for intepreting the results of X-ray studies of active carbons. Use was made here of the relation $\log I = f(\varphi^2)$ shown in Fig. 8.6, which is linear over the range of angles 0.0175–0.0524 rad. Equation (8.8) also enables determination of the total micropore volume V_{mi}:

$$V_{mi} = (NV) \approx \frac{I_{\varphi=0}}{R^3} \tag{8.9}$$

where V is the volumes of the pores.

The experimental results obtained also make it possible to estimate the number of micropores N per unit mass of the adsorbent. One should note that equation (8.8) refers, in principle, to monodisperse systems for which the distance between the inhomogeneities occurring in the structure of the solid are much greater than the size of these inhomo-

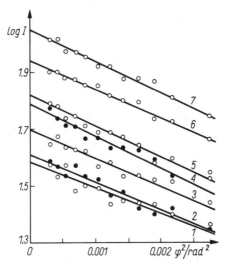

Fig. 8.6 Plots of the relation $\log I = f(\varphi^2)$ for a range of active carbons obtained from carbonized saccharose and activated with carbon dioxide to degrees of burn-off: 1 — 0%, 2 — 3.7%, 3 — 6.8%, 4 — 10%, 5 — 15.2%, 6 — 25.1%, 7 — 35,1% (after Dubinin et al. [20, 21]).

geneities. In the case of normal technological porous materials (the majority of active carbons) we are dealing with polydisperse systems with a continuous distribution of inhomogeneities (pores) depending on their size. Thus it becomes necessary to use more developed equations for describing the dependence of the scattering intensity on angle φ.

Important information on the nature of the porous structure can be obtained in another manner. The authors of papers [22, 23] assumed as the basis of their considerations the analysis of the division of a straight line passing through a porous solid into sections passing through the solid phase and sections (denoted as l) passing through the pore voids. The distribution of the sections l according to their length in the form of a function $g(l)$ was related to the intensity of X-ray scattering at small angles. If the pore distribution is not too dense, the relationship $g(l)$ is a function of the distribution of the lengths of the sections l averaged over all directions and pores. In the case of dense packing of the pores, the problem becomes much more complicated.

Function $g(l)$ has one more important feature i.e. it enables conclusions as to the shape of the pores (inhomogeneities). If the shapes are irregular (sharp edges), the value of function $g(l)$ increases with a decrease of the variable l, whereas if the pores have a circular cross-section, then the value of $g(l)$ decreases as l tends to zero. As an example of these trends we can cite [23] the observed significant increase of $g(l)$ for small values of $l(< 1$ nm) after samples of polyacrylonitrile-based carbon (not undergoing graphitization) had been heated at 2000° C and 2800° C. The increase of the pyrolysis temperature produced a change in the shape of the pores to a more regular one (close to circular with a smaller number of edges).

The method of small-angle scattering of X-rays, in addition to applications to the porous structure of active carbons (especially as regards micropores and also, to some extent, mesopores of smaller diameters) also provides the basis for other important information. For instance, by comparing experimental adsorption data regarding the volumes of various kinds of pores with the results of X-ray studies, we can evaluate the proportion of closed pores (the X-ray small-angle scattering method gives us the sum of the volumes of the open pores in contact with the surface of the carbon particles and of the closed pores).

The X-ray small-angle scattering method is finding ever-wider application as a tool in the study of changes of the porous structure of carbons due to thermal treatment in the activation process, treatment with acids, etc. It deserves stressing, however, that the interpretation of the experimental data obtained from X-ray small-angle scattering tests present some difficulties, for instance, related to the estimation of the effect of intermolecular interference when the packing of the pores (inhomogeneities) in the solid becomes very dense.

8.5 ELECTRON MICROSCOPY

Electron microscopy is a direct method of testing the porous structure of solids (including active carbons and other carbonaceous materials) [24, 25]. The main advantage of this method over, e.g., adsorption methods, is the possibility of determining directly the shape and size of the pores (especially mesopores) which in turn allows us to estimate the usefulness and accuracy of other methods, to determine the most appropriate values of the parameters used in these methods, and to establish the most reliable range of applications for them. It is important therefore to use electron microscopy alongside other methods. In many cases, for instance, not only is information obtained about the shape and size of the mesopores, but also yielding differential curves of pore volume distribution according to radii (often in good agreement with results obtained from adsorption tests or mercury porosimetry). However, electron microscopy also has certain drawbacks. Among these the more important are (i) the poor reliability of results obtained for of adsorbents with a wide distribution of pore sizes and (ii) the painstaking preparation of the samples for testing. The definition of a specification for the test solid presents an important problem. The preparation of the adsorbent samples is conducted by many different methods. One of them consists in obtaining a microsection of the adsorbent and placing it in a transparent polymer [26, 27]. Another procedure, the so-called replica method, consists in coating the adsorbent surface with a collodion solution [28] which, after drying, yields a three-dimensional replica of the surface. After peeling off the film a thin layer of carbon, silicon or gold is sprayed under vacuum onto its surface in order to increase the contrast. However, in view of the very complex pattern of the active carbon particle surface, it is difficult to obtain a faithful replica. Despite these drawbacks, electron microscopy is used quite frequently as a complementary method alongside adsorption and porosimetric tests.

References

[1] Gregg, S. J., Sing, K. S. W., *Adsorption. Surface Area and Porosity*, 2nd Ed. Academic Press, London, 1982.
[2] McBain, J. W., Bakr, A. M., *J. Am. Chem. Soc.* **48**, 690 (1926).
[3] Cameron, A. E., Reyerson, L. H., *J. Phys. Chem.* **39**, 169, 181 (1935).
[4] Urquart, A. K., Williams, A. M., *J. Text. Inst.* **15**, 43 (1924).
[5] Dreving, B. P., Kiselev, A. V., Él'tekov, Yu. A, *Dokl. Akad. Nauk SSSR* **86**, 349(1952).
[6] Él'tekov, Yu. A., *Kombinirovannaya adsorbtsionnaya ustanovka*. (Combined Adsorption Apparatus), CITEIN 1961.
[7] Lasoń, M., Żyła, M., *Chem. Anal.* **8**, 279 (1963).
[8] Ciembroniewicz, A., Lasoń, M., *Roczniki Chem.* **46**, 703 (1972).
[9] Gregg, S. J., *J. Chem. Soc.* 1438 (1955).
[10] Łukaszewicz, J., *Wiadomości Chem.* **33**, 669 (1979).
[11] Cutting, P. A., Parkyns, N. D., *J. Vacuum Sci. Technol.* **13**, 543 (1976).
[12] Kienle (von), H., Bäder, E., *Aktivkohle und ihre industrielle Anwendung*, Ferdinand Enke Verlag, Stuttgart 1980.
[13] Paryjczak, T., *Gas Chromatography in the Study of Adsorption and Catalysis*, Ellis Horwood, Chichester, PWN Warsaw 1985.
[14] Czapliński, A., *Archiwum Górnictwa* **10**, 239 (1965).
[15] Czapliński, A., Lasoń, M., *ibid.* **10**, 53 (1965).
[16] Dziubalski, R., Korta, A., Smarzowski, J., *Koks, Smoła, Gaz* **10**, 296 (1979).
[17] Dziubalski, R., Korta, A., Smarzowski, J., *ibid.* **11**, 308 (1979).
[18] Smišek, M., Černý, S., *Active Carbon*, Elsevier, Amsterdam-London-New York 1970.
[19] Guinier, A., Fournet, G., *Small-Angle Scattering of X-Rays*, J. Wiley, NewYork-London 1955.
[20] Dubinin M. M., Plavnik, G. M., Zaverina, E. D., *Carbon* **2**, 261 (1964).
[21] Plavnik, G. M., Dubinin, M. M., *Izv. Akad. Nauk SSSR, Ser. Khim.* 628 (1966).
[22] Mering, J., Schoubar, D., *J. Appl. Crystsllogr.* **1**, 153 (1968).
[23] Perret, R., Rouland, W., *ibid.* **1**, 308 (1968).
[24] Fryer, J. R., *Characterisation of porous solids. Proceedings of Symposium in Neuchatel 1978*, London SCI 1979, p. 41.
[25] Evans, M., Marsh, H., *Characterisation of porous solids, Proceedings of Symposium in Neuchatel 1978*, London SCI 1979, p. 53.
[26] Clinton, D., Kaye, G., *Carbon* **2**, 341 (1965).
[27] Kaye, G., *ibid.* **2**, 413 (1965).
[28] Lamond, T. G., Marsh, H., *ibid.* **1**, 293 (1964).

CHAPTER 9
Applications of Active Carbon

9.1 ADSORPTION FROM THE GAS PHASE

9.1.1 General Remarks

A significant fraction of manufactured active carbon (in the developed countries about one fifth of the total production) is utilized in various branches of industry as an adsorbent in gas phase adsorption. One can specify several main areas of application of the adsorption of gases on active carbon (usually in granulated form) such as: removal of substances hazardous to humans and their environment from industrial exhaust gases (pollution control of the atmosphere), recovery of valuable components from industrial gases, separation of gas mixtures, and purification of process gases from undesirable pollutants prior to use. Depending on the adsorbate, either pure commercial active carbon or active carbon impregnated with one or more catalysts are used. While in the first case one is dealing with the phenomenon of gas adsorption (often only physical adsorption), the process in the second case is more complex and involves both adsorption and catalytic decomposition of the adsorbate.

We shall consider in turn the various areas of application of active carbon of major significance.

9.1.2 Purification of Industrial Exhaust Gases, Recovery of Valuable Components, Separation of Gas Mixtures

One of the major industrial applications of the adsorption of gases on active carbon is the removal from exhaust gases of toxic components containing sulphur such as sulphur dioxide, hydrogen sulphide, carbon disulphide and organosulphur compounds.

The main sources of sulphur dioxide emission to the atmosphere are coal-fired power stations, both ferrous and non-ferrous metallurgical industries, and chemical (mainly sulphuric acid plants) as well as the petrochemical-based industries. Sulphur contained in fuels and metal ores yields primarily sulphur dioxide during their combustion or processing. Preliminary desulphurization of fuels (e.g. sulphur-rich coal) and metal ores is only partially effective. The annual emission of sulphur dioxide to the atmosphere in highly industrialized countries amounts to several dozen million tons. Its recovery is thus not only an ecological problem but also a commercial one, since more and more often the sulphur dioxide recovered is used in the production of sulphuric acid.

Among the many methods of recovering and neutralizing sulphur dioxide from exhaust gases, its adsorption on active carbon plays an ever-greater role. For gases that do not contain oxygen and water vapour, it is a purely physical process (the heat of adsorption amounts to about 44 kJ mol^{-1}). Most often, however, exhaust gases contain oxygen and water vapour along with other components. In this case the physical adsorption of SO_2 on active carbon is accompanied by its catalytic oxidation to sulphur trioxide, the subsequent formation of sulphuric acid (in the presence of water) and its partial dilution. In this case sulphur is trapped on the active carbon surface in three forms: as physically adsorbed sulphur dioxide, as a solution of sulphuric acid and as sulphur compounds bound to the surface. The proportion of these forms depends on the temperature and composition of exhaust gases. While higher temperatures favour an increase in the amount of sulphuric acid, at lower temperatures the ratio of sulphuric acid to the physically-adsorbed sulphur dioxide becomes comparable. The physically-adsorbed sulphur dioxide can be removed from the active carbon either by applying a vacuum or by purging with a gas at the adsorption temperature. The adsorbed sulphuric acid can be removed by washing with water. To facilitate extraction, active carbons with a moderately developed microporous structure are normally used. Modern plants used for purifying flue gases operate by several different methods. These differ as regards the mode of operation of the adsorbers (stationary or fluidized bed), the regeneration of the adsorbent, as well as in many details of design. Most of these methods were developed in the sixties or seventies. Among the most commonly used in practice are (i) the Japanese methods — Hitachi and Sumitomo, (ii) the German methods — the Lurgi, Stratman, Bergbau-Forschung processes and the Reinluft method, and (iii) a method developed in the USA, the Westvaco (West Virginia Pulp and Paper Co.) method. A detailed description of the plants used for purification of gases by these various methods together with the schemes and conditions of operation can be found in the literature [1–3]. Although varying in many details, all these methods have certain important features in common. Therefore, in order to give a general illustration of the principle of gas purification processes, we shall rely on the Hitachi method.

Flue gases containing SO_2 are first fed to dust catchers and then, by means of a blower, to a section of several adsorbers filled with active carbon and constituting a part of a battery operating in a parallel system. The process of purification is conducted at temperatures exceeding 100°C. Simultaneously another section of the battery of adsorbers is washed, firstly with concentrated, and then with dilute sulphuric acid, and finally with water. The liquid obtained from the washings is a 20% aqueous solution of sulphuric acid. This solution is then de-oiled in an oil separator and concentrated to 70% by evaporation. After passing through a cooler and a centrifuge the acid is pumped to storage tanks. While adsorption is proceeding in one section of the battery and the washing cycle in another, a part of the contaminated flue gas is passed through the third section containing washed but still wet (up to 50% moisture) carbon. Drying of the carbon and partial purification of the flue gas then takes place at the same time. This gas is further combined with the stream of purified gas and vented to the stack. The approximate durations of the individual steps are as follows: purification 30 h, washing 20 h and drying 10 h. The degree of purification of the flue gases achieved usually exceeds 90%.

Another serious problem related to the protection of the natural environment against noxious sulphur-based pollutants is the purification of the waste gases emitted to the atmosphere by the rayon industry. The main toxic components of these gases are usually hydrogen sulphide (60–2000 mg m^{-3}) and carbon disulphide (300–8000 mg m^{-3}). The effective removal of these substances is important since, on the one hand, contamination of the atmosphere is prevented and, on the other hand, considerable economic benefits are achieved through the recovery of valuable raw materials. Carbon disulphide is usually removed (and recovered) by applying active carbon, while hydrogen sulphide is removed in some systems by absorption in liquid media. However, this technique does not ensure complete removal of H_2S. In the last dozen or so years, several different modifications of a new method have been developed in which both hydrogen sulphide and carbon disulphide are removed simultaneously from waste gases by means of active carbon.

Of these various modifications which differ primarily with regard to the use of the carbon adsorbent, we shall present those of special interest. In the first method, adsorption of both adsorbates is carried out at 20–50°C in two stationary beds of different kinds of active carbon. In one bed, consisting of a macroporous carbon usually impregnated with iodine, hydrogen sulphide is adsorbed and oxidized to sulphur which is deposited in pores of greater effective radii (chiefly mesopores). In the second bed, consisting of microporous carbon, adsorption of carbon disulphide takes place. After the adsorbent is taken out of operation, the worn-out active carbon is regenerated. During regeneration carbon disulphide is removed with a stream of superheated steam and sulphur is extracted with liquid carbon disulphide. Since hydrogen sulphide may undergo partial oxidation to sulphuric acid on the surface of active carbon, gaseous ammonia (introduced periodically or continuously with the waste gases) is usually used for its neutralization. The ammonium salts generated are easily removed with water in the regeneration cycle. In another method a bed of one kind of active carbon (macroporous carbon with a low iron content) is utilized. Hydrogen sulphide, contained in the waste gases fed to the adsorber, is oxidized in the front layer of the stationary carbon bed. Ammonia is used to neutralize the small quantities of sulphuric acid that might be generated. At the same time physical adsorption of carbon disulphide proceeds. The regeneration of the carbon bed is conducted in a multistep cycle when the particular adsorbates are recovered from the carbon. The Sulfosorbon process provides an example of the former procedure and the Thiocarb or Pintch-Bamag processes, the operating parameters of which are described in the literature [1–3], as an example of the latter.

The process of catalytic oxidation of hydrogen sulphide on active carbon finds application not only in the purification of waste gases in the chemical fibre industry but also in the desulphurization of various industrial gases such as water gas or coke oven gas [1]. The considerable quantities of sulphur recovered in this process constitute a valuable raw material. Since catalytic oxidation of hydrogen sulphide on active carbon is highly exothermic, it is usually applied when the concentration of hydrogen sulphide does not exceed 5 g m^{-3}. Sometimes, however, this process is conducted even if the concentration is twice as large. In such a case the temperature in the carbon bed increases to almost 100°C. Some ammonia is usually added to the stream of gases to be purified.

Ammonium sulphate is then obtained alongside the main product, and, depending on the composition of the initial gas, also ammonium carbonate. To regenerate active carbon saturated with sulphur, an aqueous solution of ammonium sulphide (capable of dissolving considerable quantities of sulphur in the form of ammonium polysulphide) is usually used. In this case sulphur is extracted from the bed successively with several solutions, firstly with a solution containing a large amount of sulphur, then with a solution of low sulphur content and finally with a fresh ammonium sulphide solution. Sulphur is then recovered by treating the polysulphide solution with steam at 125°C. Ammonium salts present in the carbon bed are removed by washing with water. In some instances, in addition to ammonium sulphide, xylene is used to remove sulphur from the carbon bed. The virtue of xylene lies in the strong temperature-dependence of the solubility of sulphur in this solvent (increase of temperature from ambient to 100°C produces a tenfold increase of the solubility of sulphur). The removal of sulphur from the carbon bed is carried out at about 100–110°C, while isolation of sulphur is achieved by its crystallization at about 30°C and separation in a centrifuge. The process of purifying gases from hydrogen sulphide is usually conducted in batteries of adsorbers, enabling their successive switching over to the cycle of adsorption combined with catalytic oxidation of hydrogen sulphide, and to the regeneration cycle involving extraction of sulphur and removal of the extracting solutions from the carbon bed.

In addition to these applications to the purification of gases and recovery of sulphur-containing materials, active carbon is also used in many other processes the aim of which is the removal of noxious components of gases or the recovery of valuable substances. The control of nitrogen oxides is another example. This is performed in various ways [2, 3]. In one case nitrogen oxides are converted at 200–300°C on active carbon with a deposited metallic catalyst and in the presence of sufficient ammonia into nitrogen and water. In another procedure, utilizing pure active carbon, nitrogen oxides are converted to NO_2 which is then absorbed.

An important application of active carbon is the recovery of organic solvents from technological or waste gases (recuperation). This process is of significance both from the commercial point of view (recovery of valuable materials), and as regards protection of the natural environment (control of air pollution). The main solvents recovered in this way are petrol, benzene, toluene, xylene, acetone, lower alcohols, diethyl ether, n-alkanes (C_6—C_7), halocarbons (mainly chloroform, carbon tetrachloride, methylene chloride, dichloroethane, chlorobenzene) and carbon disulphide. The concentrations of organic solvent vapours in waste gases are usually small, ranging from one to a dozen or so g m^{-3} of gas. The application of active carbon (usually in a stationary bed) for trapping them is, therefore, effective — the concentration of solvent in purified gas does not exceed 0.5 g m^{-3}. The recovery plants working in various countries are often largely automated and, despite some differences, show considerable similarity as regards the general principles of their operation. In view of the character of the adsorption process, the temperature of the gases entering the plant is kept as low as possible (cooling sometimes being applied) — it should not exceed 40–50°C. The flow of gas often proceeds from the bottom of the bed upwards. Usually batteries of adsorbers operating in parallel are used. In the desorption step, superheated steam is passed through the bed in the direction contrary to that of gas flow in the adsorption step. Steam, together with the solvent vapours, is cooled and condensed and the solvent layer is separated from water in a tap funnel. After regeneration with steam, active carbon is dried and cooled with atmospheric air before its subsequent use. The use of recovery plants will undoubtedly spread in view of the fact that ever-greater amounts of various solvents are used in numerous industrial processes. More details on the subject can be found in the literature [2, 4].

A different field of application of active carbon is the control of disagreeable odours in air due to the presence of often very small quantities of such substances as, e.g., phenol, chlorophenol, pyridine, diethyl ether or mercaptans. Various filters containing active carbon are used for the purpose [3].

Active carbon is also used in many instances for the separation of gaseous mixtures into single components or groups of components. Thus we can mention the recovery of benzene from city gas or gasoline, of propane and butane from natural gas, as well as the separation of gases from the low pressure Fischer–Tropsch synthesis [2, 4].

In nuclear power plants active carbon is applied in the control of radioactive ^{131}I and ^{133}I isotopes produced in nuclear fission reactions and occurring in air both in elemental form and as chemical compounds, e.g. methyl iodide. Carbon impregnated with potassium iodide is usually used in the filtering devices, since a high degree of air purification must be ensured even under conditions of high humidity. Active carbons are also used in nuclear power plants for control of radioactive noble gases (e.g. ^{85}Kr or short-lived krypton and xenon isotopes) in air [2].

New applications of adsorption on active carbon have been created by the development of carbon molecular sieves. Methods of their fabrication as well as the raw materials used (i.e. anthracite, oxidized hard coal, plastics) are described in [3]. When compared with typical active carbons, carbon molecular sieves are characterized by their narrow distribution of micropore volumes. The major part of the micropores have effective diameters comparable to the size of the adsorbed molecules (usually in the range of 0.5–1.0 nm). Carbon molecular sieves have found many applications, e.g. in the separation of air into its components: oxygen and nitrogen. Depending on their parameters, gases of different degrees of enrichment are obtained [3, 5, 6].

Although these do comprise an exhaustive list of all the possibilities, the applications discussed above of active carbon in gas phase adsorption process are currently the most important. These applications will certainly expand as new technological solutions are implemented in various industries.

9.1.3 Applications of Impregnated Active Carbons for Protection of the Upper Respiratory Tract against Toxic Substances

Among the system composed of active carbon with deposited substances that play an important role as catalysts of many chemical reactions proceeding in the gas phase, the so-called impregnated active carbons deserve special attention. They enable the protection of man's upper respiratory tracts against harmful substances that may penetrate into the lungs together with air. In this case the active carbon plays the role of a carrier on which various substances are deposited from the liquid phase. This process is known as impregnation and the carbons as impregnated active carbons. Such carbons were applied for the first time during World War I for protection of the upper respiratory tracts of soldiers against warfare gases.

The substances used for impregnation are compounds of the following metals: copper, chromium, silver, potassium, sodium, zinc, cobalt, manganese, vanadium and molybdenum, as well as certain organic compounds such as pyridine or aromatic amines. Depending on the adsorbate, the process of adsorption of toxic gases on the surface of impregnated active carbon is accompanied by the following phenomena [4]:

(1) physical adsorption (e.g. chloropicrin),
(2) chemisorption, i.e. reaction with the components of the impregnate: neutralization, hydrolysis or formation of complexes (phosgene, cyanogen chloride),
(3) catalytic reactions such as oxidation (arsines).

Suitably selected impregnants are almost universal for the preparation of adsorbents trapping various toxic substances. However, the composition of the impregnants is usually a laboratory secret and is rarely published. Table 9.1 presents examples of reactions of some toxic substances on impregnated active carbons as catalysts [2, 4, 7].

Copper- and chromium-impregnated charcoals (the so-called whetleryts after the names of J. C. Whetzel and E. W. Fuller) are good adsorbents for the chemisorption of toxic gases and are used to adsorb cyanogen chloride, hydrogen cyanide and arsines. It is stressed in the patent held by Wilson and Whetzel [8] that active carbon impregnated with copper and silver compounds is particularly useful for purifying air from such gases as chlorine, arsines, phosgene, hydrogen cyanide, carbon monoxide, etc. Addition of zinc oxide activates these adsorbents towards acidic gases such as phosgene and diphosgene.

The development of chemical warfare introduced new kinds of war gases and stimulated the search for new effective adsorbents. The main

Table 9.1
Reactions proceeding on the surface of impregnated active carbon used as a catalyst (after von Kienle and Bäder [2], Smíšek and Černý [4] and Jankowska and Choma [7])

Impregnant component	Effective form of impregnant	Function of impregnant	Stoichiometric equation
Copper	Cu_2O	reactant	$Cu_2O + 2HCN \rightarrow 2CuCN + H_2O$
	CuO	reactant	$CuO + COCl_2 \rightarrow CuCl_2 + CO_2$
Zinc	Na_2ZnO_2	catalyst	$2AsH_3 + 3O_2 \rightarrow As_2O_3 + 3H_2O$
	ZnO	reactant	$ZnO + 2HCN \rightarrow Zn(CN)_2 + H_2O$
	ZnO	reactant	$ZnO + COCl_2 \rightarrow ZnCl_2 + CO_2$
Silver	Ag, Ag_2O	catalyst	$2AsH_3 + 3O_2 \rightarrow As_2O_3 + 3H_2O$
Copper	$CuSO_4 \cdot 5H_2O$	reactant	$CuSO_4 \cdot 5H_2O + 4NH_3 \rightarrow$ $\rightarrow [Cu(NH_3)_4]SO_4 + 5H_2O$
Chromium	$CuCrO_4 \cdot NH_3 \cdot 5H_2O$	catalyst	$ClCN + 2H_2O \rightarrow CO_2 + NH_4Cl$
	$CuCrO_4$	catalyst	$ClCN + 2H_2O \rightarrow CO_2 + NH_4Cl$
Pyridine	C_5H_5N	reactant	$C_5H_5N + ClCN + H_2O \rightarrow CHOCH =$ $= CH—CH = CHNHCN + HCl$

problem was the selection of appropriate impregnants. Currently in industrial practice, active carbon is impregnated with a mixture of the following solutions: tetramminocopper(II) (cuprammonium) carbonate, ammonium chromate and silver salts. Active carbon impregnated with these substances is called chromium-copper adsorbent. One typical composition of the impregnating solution calculated per 100 kg of active carbon is as follows [9]:

basic copper carbonate	6 kg
ammonium carbonate	5 kg
ammonia liquor (25%)	10 l
potassium dichromate	3 kg
silver nitrate	0.17 kg
water	59.5 l

The composition of the impregnant on the surface of the chromium-copper adsorbent is not yet precisely known. Serious experimental difficulties are encountered in this area. It has been found for example [10] that on the surface of chromium-copper active carbon adsorbents there exists a mixture of catalysts of composition: ca. 50% of copper chromates of the formulae — $CuCrO_4 \cdot NH_3 \cdot H_2O$ and $CuCrO_4 \cdot 2CuO \cdot 2H_2O$, and ca. 50% of highly deformed (and amorphous from the X-ray point of view) chromium and copper oxides. In standard conditions, protection against cyanogen chloride and hydrogen cyanide is best when the chromate content is greatest, whereas protection against arsine is highest when the oxide content is largest. The chromium-copper adsorbent should be heat-treated at 170–190°C to obtain a product exhibiting good adsorption and catalytic properties towards hydrogen cyanide, cyanogen chloride and arsine.

In view of its practical importance, significant attention is being paid to the adsorption of cyanogen chloride. This process is applied in warfare as a standard test reaction for the evaluation of adsorbents used for protection of the upper respiratory tract. It has been shown experimentally that active carbon impregnated with solutions containing such ions as Cu^{2+}, CrO_4^{2-}, Ag^+, NH_4^+, CO_3^{2-} is a very good adsorbent for cyanogen chloride. If a mixture of this gas and air is passed through a bed of active carbon impregnated with substances containing the above ions, then cyanogen chloride is trapped irreversibly in the layer. However, if pure (i.e. not impregnated) carbon is used, the process of cyanogen chloride adsorption is partly reversible, and some of the gas passes through the bed, so its concentration in the air beyond the carbon bed soon reaches its initial value.

Table 9.2

Parameters of porous structure of active carbon both impregnated and not impregnated with copper and chromium compounds (after Choma [11])

Parameter	Active carbon	
	not impregnated	impregnated
Volume of micropores $V_{mi}/\text{cm}^3 \text{ g}^{-1}$	0.329	0.206
Volume of mesopores $V_{me}/\text{cm}^3 \text{ g}^{-1}$	0.288	0.187
Surface of mesopores $S_{me}/\text{m}^2 \text{ g}^{-1}$	150	100
Volume of macropores $V_{ma}/\text{cm}^3 \text{ g}^{-1}$	0.523	0.405
Surface of macropores $S_{ma}/\text{m}^2 \text{ g}^{-1}$	0.05	0.04
Dunin–Radushkevich equation coefficient $W_o/\text{cm}^3 \text{ g}^{-1}$	0.374	0.230
$10^6 B/\text{K}^{-2}$	0.672	0.553

It is noteworthy that impregnation of the active carbon with catalysts causes its surface properties to deteriorate. A significant decrease in the parameters characterizing both micropores (volume) and mesopores (volume, specific surface area) is observed. This problem is illustrated in Table 9.2. One may surmise that the impregnants are deposited mainly in the mesopores, partly in the macropores, and sometimes block the entrance to micropores. This is the reason why for physically-adsorbed adsorbates, impregnation impairs the properties of the adsorbent. Adsorption of cyanogen chloride on non-impregnated active carbon is presumably of a physical nature and experiments have shown that its physical adsorption is poor. Its adsorption on impregnated carbon is a chemisorption process where the impregnants play undoubtedly a leading role. It is assumed that the adsorbed cyanogen chloride interacts with copper and chromium cations present on the surface. On the basis of the literature data and the present authors' results, the following conclusions may be drawn:

1. The impregnants deposited on the surface of active carbon play a crucial role, since their presence makes the adsorption of cyanogen chloride, phosgene, organophosphorus compounds, etc. an irreversible process. They do not improve the adsorption properties of the adsorbent (in terms of physical adsorption), but on the contrary, these properties are impaired.

2. The irreversibility of adsorption of toxic substances is related to the course of both catalytic and noncatalytic reactions as well as the desorption of the reaction products.

3. The presence of water adsorbed on the surface of active carbon and taking part in these various processes plays a significant role.

The following reactions are likely to take place on the surface of the adsorbent in the case of cyanogen chloride. It is known that under the influence of water vapour, cyanogen chloride undergoes a slow hydrolysis reaction with the formation of hydrogen chloride and cyanic acid according to the equation:

$$ClCN + H_2O \rightarrow HCl + HNCO \tag{9.1}$$

It is believed that copper chromate, present on the carbon surface, largely accelerates reaction (9.1) as well as another reaction, namely that of cyanic acid with water:

$$HNCO + H_2O \rightarrow CO_2 + NH_3 \tag{9.2}$$

Hydrogen chloride, the second product of reaction (9.1), reacts with copper oxide and with chromium (VI) compounds present on the surface according to the equations:

$$2HCl + CuO \rightarrow CuCl_2 + H_2O \tag{9.3}$$

$$6HCl + 2Cr(VI) \rightarrow 3Cl_2 + 2Cr(III) + 6H^+ \tag{9.4}$$

Since cyanogen chloride is also trapped by the absorbent which has not been exposed to water vapour, the following oxidation reaction also has to be considered:

$$2ClCN + 4O \rightarrow 2CO_2 + N_2 + Cl_2 \tag{9.5}$$

There has been a proposal that part of the cyanogen chloride undergoes hydrolysis according to the reaction:

$$ClCN + H_2O \rightarrow HOCl + HCN \tag{9.6}$$

The following reactions have been proposed for hydrogen cyanide trapped on the surface of the adsorbent [12]:

$$2HCN + CuO \rightarrow Cu(CN)_2 + H_2O \tag{9.7}$$

$$2Cu(CN)_2 \rightarrow 2Cu(CN) + (CN)_2 \tag{9.8}$$

The following hypothetical mechanism is assumed for the overall process of trapping toxic substances on the chromium-copper active carbon adsorbent [13–15]. Molecules of toxic compound are adsorbend both on the free parts of the surface and on the parts on which ions of the impregnants have been deposited. Therefore physical adsorption and chemisorption take place simultaneously. The molecules of cyanogen chloride bonded to the ions of the impregnant become chemically active and may, for instance, interact with the physically-adsorbed water molecules yielding carbon dioxide and ammonia as final products (reactions 9.1 and 9.2).

Perhaps the process leads to the regeneration of the part of the surface on which the ions of impregnants are deposited, so this surface may once again interact with cyanogen chloride molecules. Cyanogen chloride adsorbed on the free parts of the surface may also react with adsorbed water but this reaction is significantly slower. Under these conditions the adsorbed molecules are desorbed faster than they are hydrolysed. They are, however, adsorbed again on areas covered with ions, so the only gas to be desorbed in the overall process is carbon dioxide.

The great interest in the mechanism of the activity of adsorbents in the decomposition reactions of toxic substances results from its wide application not only in military technology but also in all branches on the chemical industry where, in emergencies, human lives are endangered by toxic gases leaking to the atmosphere.

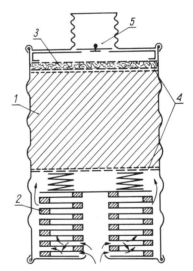

Fig. 9.1 Diagram of a respirator filter: 1 — active carbon layer, 2 — smoke filter, 3 — cotton wool, 4 — wire mesh, 5 — breathing valve (after Smišek and Černý [4]).

Additionally, as it has already been mentioned, carbon adsorbents, sometimes further modified, are utilized in almost all armies in devices designed to protect the upper respiratory tract such as filtration gas-masks, filtering-ventilation devices for shelters, and armoured fighting or transportation vehicles.

The diagram of a gas-mask filtering canister is shown in Fig. 9.1.

9.2 ADSORPTION FROM THE LIQUID PHASE

9.2.1 Food Industry

Adsorption on active carbon from the liquid phase has found applications in the food industry for many years in various areas of production and processing of food [2]. In each case the function of the active carbon is to remove unwanted components from solutions, which results in significant improvement (purification) of the main dissolved component, i.e. the food product. The application of active carbons in this field is continuously increasing. This is due both to the growth of production in those areas where the application of active carbons is already well-established, and to the extension of their use to food technologies in which they were not previously utilized.

Let us discuss in some detail several more important applications of active carbon in the food industry.

Application of active carbon in the sugar industry. Solid adsorbents were already in use at the beginning of the nineteenth century for the purification of syrups. Initially, for quite a long time, bone charcoal was used, but, at the turn of the twentieth century, this was gradually replaced by active carbon. Great Britain and then Germany were the first to introduce active carbon into the sugar industry. The large increase of demand in this industry was one of the principal reasons for the fast increase of active carbon production on a commercial scale. The first active carbon plants were built in 1911. These were in Raciborz (Silesia) and in Stockerau near Vienna, and at about the same time was built the 'Norit' plant by the Norit Company of Amsterdam.

Decolorization of sugar syrup by means of a carbon adsorbent is the last step of sugar purification. Among discolouring substances that, must necessarily be removed are, principally, (i) melanoydines, i.e. nitrogen-containing dyes produced during sugar reduction with amine substances, (ii) caramels, i.e. nitrogen-free dyes produced as a result of partial thermal decomposition of sugars containing phenol, polyphenol and quinoid groups, (iii) iron complexes and small amounts of other substances. The substances listed above occur both in undissociated and dissociated forms (aionic forms prevailing). Note that the most intense colour is due to substances of molecular mass 8000–15 000 occurring in colloidal form. Even a small improvement of the purity of the solution by the application of active carbon not only effectively removes its colour but also improves the properties of the product from the technological point of view. The removal of surfactants and colloidal substances increases the surface tension of the syrup with a simultaneous decrease of its viscosity. As a result the rate of crystallization increases and the separation of sugar crystals in the centrifuge are improved. Mineral impurities (salts of certain metals) are present in the sugar syrup along with organic compounds. The removal of these impurities from the syrup is also important as they increases the amount of molasses (due to the binding of part of saccharose).

In the food industry, active carbon is used today in several forms: powder, granules, decolorizing ion exchangers (in some cases), and to an ever-decreasing extent, bone charcoal. While in some sugar plants only one form of adsorbent is used, in others better results are achieved by combining different adsorbents at different stages of production. Decolorization with powdered active carbon may be carried out by two different methods (a third being their combination).

The first method is a batch process. It consists of filling the containers with sugar syrup and a given amount of carbon (5–10 kg m^{-3} of syrup). The resulting suspension is maintained at 80–90°C for about 20 min, which is sufficient to achieve adsorption equilibrium. Next the suspension is pumped to a filter press. During filtration the thickness of the carbon bed on the filtering element increases to about 30 mm with a resulting increase of flow resistance from 2×10^5 to 4×10^5 Pa. About 600 kg of carbon is collected in 1 m^3 of filter volume. When the press is filled up, the active carbon is rinsed with water (5–10 fold mass excess with respect to the carbon) and removed. The decolorization process is often carried out in two steps (especially) for cane-sugar.

A simplified diagram of the plant discussed, operating in a counter-current mode is presented in Fig. 9.2. The amount of carbon used in such a system is much smaller. In step I the crude syrup is brought into contact in tank 1 with the carbon from step II, leaving tank 2, where already initially-decolorized syrup is treated with fresh carbon. The spent carbon (after its utilization for the initial purification of syrup

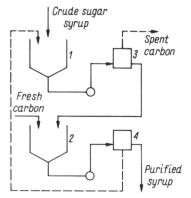

Fig. 9.2 Diagram of a two-step sugar syrup purification plant: 1,2 — contactors, 3, 4 — filter presses.

(in step I) is removed from the system. For economic reasons active carbon is regenerated in larger plants only. Initially it is boiled in dilute (1–4%) hydrochloric acid. After the acid is removed by washing with water, the active carbon is boiled in a 1% solution of either sodium hydroxide or sodium carbonate in order to remove the adsorbed dyes. After drying, the carbon is heat-treated at 500–600°C when the organic substances adsorbed during the decolorization of syrup are decomposed. The resulting amorphous carbon is deposited in the pores suppressing by 10–20% the adsorptive capacity of the adsorbent. The losses of carbon in the regeneration process amounts to about 5% and they are made up by adding fresh carbon.

The second method, in which powdered carbon is used, consists of passing syrup through a 10–15 mm thick carbon bed that is deposited on a filtering fabric covered with diatomite. Various types of filters are used in the filtration process. Because of the continuous reduction of the decolorizing capacity of the bed during filtration, systems of multiple filters are usually used which ensure a constant degree of syrup purification. The spent carbon may be regenerated.

Note that each of the methods discussed above requires the application of a suitable carbon of specified properties. In the first method it must be a carbon of good filtering properties (the flow rates are here about ten times higher than in the continuous method) while its decolorizing ability may be lower (for instance Norit Standard). In the second method (continuous filtering through the carbon layer) the application of a carbon of better decolorizing ability (Carboraffin, for instance) is recommended. Here the filtering properties of the carbon are less important. Another major difference between these two methods relates to the consumption of carbon, which in the first case amounts to 0.3—1.0% of refined sugar, whereas in the second case it could be ten times smaller, i.e. 0.05–0.1%, but the quality of the carbon must be higher. The selection of one of these methods (or their combination) depends on the size of the plant and should be based on economic grounds.

Granulated active carbon is also often used in sugar syrup purification processes. In this case the decolorizing process may be conducted by several methods, namely the stationary bed method, the fluidized bed method, and the counter-current continuous method.

In the first method a solution (heated to about 80°C) is passed downwards through a stationary bed of adsorbent at a linear velocity of 1.5–2.5 m h^{-1}. This velocity ensures a contact time between the syrup and adsorbent of about 3–5 h, whereas the total working time of the bed amounts to about 20–30 days. A system of adsorption columns is usually applied enabling simultaneous decolorization in one section of the plant, washing the adsorbent in another, and its exchange for fresh (regenerated) carbon in the third one. The regeneration of the spent carbon is conducted in the cycle: washing, drying and heat-treatment in a slightly oxidizing atmosphere at 1000–1100°C. Activation with steam is also possible here. The adsorption properties of the regenerated active carbon are usually decreased by 5–10%.

In the second method certain (relatively small) amounts of spent adsorbent are removed from the bottom of the filter and a fresh portion of carbon is supplied to the top of a bed. The bed is therefore gradually displaced in the counter-current direction with respect to the continuous flow of the syrup.

In the third and final method (continuous counter-current mode) the solution to be decolorized flows upwards through the layer of granulated adsorbent moving down the adsorption column. The flow rate of the syrup is adjusted so (linear velocity of about 3.5 m h^{-1}) as to extend the bed but not allow it to become fluidized. The advantage of such a procedure lies in the exposure of the entire surface of each pellet in the same way to the flowing solution. In the non-extended bed this would be impossible since part of the pellet surface area would be blocked due to the mutual contact when close-packed. In contrast, in a fluidized bed, mixing of the grains of different degrees of usage takes place, which affects in a negative sense the adsorptive capacity of the bed. The method described above, first applied for decolorizing syrup in a sugar plant in Woodland (USA), is becoming more and more popular because of its advantages. The plant usually consists of several adsorption columns, a column for the recovery of sugar from the spent carbon, a hydraulic transportation system, a system for removing water from carbon (cyclone), and a rotating regeneration furnace (operating at *ca.* 900°C). Regenerated carbon is continuously returned to the decolorizing process, which can then be conducted in a continuous and automated regime. Due to the very good utilization of the adsorptive capacity of the bed, the continuous counter-current method is very economical. However, a slightly lower decolorization efficiency, as compared with methods using powdered carbon, is a certain disadvantage of this technique.

Different combinations of active carbon with ion-exchangers are also used for the decolorization [16]. Carbon plays a protective role, preventing over-rapid exhaustion of the ion-exchanging bed.

Currently, many different methods of syrup purification with active carbon are used in the sugar industry. This variety results from the differences in the size and age of the plants as well as from other economic factors.

Application of active carbon for decolorizing oils and fats. Mixtures of active carbon with bleaching earth rather than pure carbon are used for decolorization (bleaching) of oils and fats [17]. The qualitative and quantitative compositions of such mixed adsorbents depend on the kind of material to be decolorized. For economic reasons, combined adsorbents with a low content of active carbon (which is virtually only an additive to the relatively cheap mineral adsorbent) are normally used. However, the active carbon plays a predominant role in such a mixed adsorbent, since even its low content contributes to a significant reduction of the mass of the bed that would be required to achieve the necessary degree of decolorization. The composition of the mixture is controlled by economic factors. In each case it is adjusted to a given oil or fat which is why decolorization of oil or fat is usually a batch process. Oil, together with the optimum amount of adsorbent, is fed to a vacuum mixer to avoid oxidation. After 10–20 minutes of vigorous stirring at 100–120°C, the adsorbent and the oil are separated on a filter press. The part of the oil remaining in the sludge is extracted with petrol or occasionally with other solvents. Since the adsorptive capacity of the bed is not completely exhausted, it may be utilised for preliminary purification of another batch of oil. The completely spent adsorbent is discarded. The conditions of the process described above cannot be applied when oil contains vitamin A which must be preserved by carrying out the process at room temperature; moreover, oil is then usually diluted with a suitable solvent, such as heptane.

Application of active carbons for improving the taste and other properties of alcoholic beverages. Active carbon finds increasing application in the production of alcoholic beverages at different stages of the process. In each case the aim is to remove undesirable components, which leads to an improvement in taste and other properties.

In the case of rectified spirit the situation is as follows. Three main fractions are collected during rectification of the crude spirit: the light fraction (low boiling aldehydes and other oxygen-containing organic compounds), the medium fraction (the spirit itself) and the heavy fraction (fusel oil). Rectified spirit always contains certain amounts of fusel oil characterized by an unpleasant taste and odour, but this is removed successfully by passing the spirit through a layer of active carbon (which has replaced the lime-tree charcoal used before).

In the production of brandy, active carbon is used mainly for the removal of acids, furfural and tannin. The application of about 5 g of powdered carbon per litre of brandy gives good results. By increasing the amount of added active carbon (to about 30 g dm^{-3}) the micro amounts of fusel oil can be partly removed and a distinct improvement of taste is acquired, but ethers, acetaldehyde and higher alcohols are not adsorbed.

In wine production, active carbons with special properties are used. They should be able to remove undesirable components originating from mildew, cork, yeast, etc. which may give an unpleasant taste and odour. At the same time active carbon is required to modify the colour of the wine in a predetermined way. However, since active carbon may also adsorb other components of wine, the adsorption method is recommended only if other means fail. A more detailed investigation on the application of active carbon in wine production was carried out by Singleton and Draper [18], who used frontal chromatography to determine the ability of several dozen different active carbons to adsorb particular substances present in wine. In view of the difficulties mentioned above it is important to carry out an analysis of the wine and to select the adsorbent best conforming to the results of this analysis. Carbon in powdered form may be added to either the wine or the must in an amount equal to about 1 g per litre of wine. This quantity may be increased or decreased depending on the type and quantity of the undesirable components. First a suspension of carbon in a small amount of wine (must) is prepared and then added to the rest of the wine (must) placed in a vat for purification. In order to promote adsorption, mechanical stirring or bubbling with a stream of air fed at the bottom of the vat is applied. After several hours the carbon is separated, first by sedimentation and then by filtration on a filter press or in a centrifuge. Despite certain disadvantages (insufficient selectivity in the removal of certain wine components) this method of the improvement of the taste and colour of the wine constitutes an important step in the manufacturing process. It may be expected that the production of new kinds of active carbon with designed properties will bring further progress in this field.

Active carbon is used in the brewing industry at different stages of the manufacturing process. It is used for the purification of water, air and carbon dioxide. It is also applied directly in the process of brewing to modify the colour and to remove the taste and odour of phenol. The adsorption method of removing the undesirable components of beer gives quite good results. The concentration of these components can be greatly reduced in this way. Usually about 20–25 g of powdered active carbon per 100 dm^3 of beer is added directly before racking. Note however, that since carbon also removes the valuable components, it should be added in the possible smallest amounts, determined in each case by laboratory tests.

9.2.2 Water and Wastewater Treatment

Purification of municipal water. The problem of municipal water conditioning is very important especially in highly developed countries and in large urban agglomerations. Rapid industrial development and the resulting increase of the amount of liquid wastes disposed of by industry brings about a continuous increase in the concentration of impurities in the sources of water supplying urban water lines. The increasing chemicalization of agriculture (e.g. fertilizers, herbicides, etc.) and of almost every other domain of life (e.g. detergents) brings similar hazards. The continuously increasing variety and amount of hazardous compounds present in our lakes, rivers and sometimes also in underground water reservoirs make the conventional means of water purification insufficient, and the introduction of new, more effective technologies becomes essential. Active carbon plays a predominant role in this area. Together with strong oxidants, especially ozone, active carbon provides a key tool in the process of removing refractory substances, including micropollutants from water, and determining its potability. The most common impurities are pesticides, detergents, aliphatic and aromatic hydrocarbons, phenols and their derivatives, carcinogenic compounds, heavy metals, viruses, etc. In many water conditioning plants active carbon is applied successfully to remove substances imparting an unpleasant taste and odour to water as well as in its dechlorination. In recent years special methods involving the use of active carbon together with ozone have been developed, for instance, as a biological bed for the nitrification of nitrogen present as ammonium compounds. Active carbon is used in various forms: powdered, granular or granulated. Many factors have to be taken into consideration when selecting one of these forms for water conditioning. Among the most important are the appropriateness of using carbon, the technological and economic possibilities, the type and efficiency of the water conditioning plant. If the technology utilizing active carbon is introduced in an already existing plant, the powder form is usually recommended. However, the decision as to which carbon should be selected should be preceded by investigations on a pilot plant scale carried out under the particular conditions of a given source of water and over a period of time long enough to allow for seasonal variation in the water composition. The results of this research should provide the basis for decision as to the amount of applied carbon required, the points at where it should be introduced in the technological process, the selection of the appropriate brand, and, in the case of powdered carbon, also the mode of preparation of the suspension. It should be stressed that often two different forms of carbon are used simultaneously.

The most common technological systems of water purification in which active carbon is used are:

(a) Stations in which adsorption is used mainly for removal of substances with an unpleasant taste and odour.

(b) Systems in which carbon is also used for removal of oxidation products resulting from the treatment of water with strong oxidants.

(c) Systems (less familiar) designed for treatment of water containing, along with other impurities, nitrogen in the form of ammonium compounds (3 g m^{-3}) subjected to oxidation to nitrates in the carbon bed.

The dosage of active carbon is a very important factor in these technologies. In the case of powdered carbon, the average dose amounts to $5-30 \text{ g m}^{-3}$, and sometimes even more. The dosage may be carried out by two different methods. The first or 'dry' dosage is used in large water conditioning plants. Here the dosing devices take the carbon directly from the storage tanks and feed it directly to the wetting device. In the 'wet' dosing process the first step consists of the preparation of a suspension containing up to 10% of carbon in a tank provided with a stirrer. This suspension is stored for a certain time before use. The devices in which granulated carbon is used are commonly known as active carbon filters, though in fact they are adsorption beds which play or may play, simultaneously, the role of filters. The beds most commonly used in water treatment stations are those with a downward flow. In contrast to sand beds, carbon beds have to be replaced relatively often. The spent carbon may be regenerated and returned to the process.

Along with direct purification, dechlorination of water deserves special mention among major applications of active carbon at different steps of the water conditioning process. This problem follows from the chlorination of water for many years to secure disinfection, removal of ammonium compounds and reduction of colour, unpleasant taste and odour. Chlorination also facilitates the removal of organic impurities (especially in the conditioning of water for industrial use). Since potable water should not contain more than 0.1 mg dm^{-3} of chlorine (in some industrial processes this limit is even as low as 0.1 mg dm^{-3}) dechlorination is necessary to remove the excess chlorine introduced at the chlorination stage. Removal of the chlorine compounds formed in the water is also necessary. Adsorption on active carbon is today the simplest and the most effective method of dechlorinating water. The mechanism of the processes taking place when chlorinated water is brought into contact with the surface of active carbon is complex. On dissolution of chlorine in water its hydrolysis proceeds according to the equation:

$$Cl_2 + H_2O \rightarrow H^+ + Cl^- + HClO \tag{9.9}$$

The equilibrium constant for this reaction at 25°C is $K = 3.94 \times 10^{-4}$ [19]. Introduction of active carbon into chlorinated water has a significant effect on this reaction. A rapid loss of chlorine is observed with the simultaneous formation of hydrogen chloride. This process is accompanied by chemisorption of an equivalent amount of oxygen on the carbon surface. The reaction may be described by the equation [20, 21]:

$$C_x + HClO \rightarrow C_xO + H^+ + Cl^- \tag{9.10}$$

or, when water contains ammonium compounds:

$$C_x + 2NHCl_2 + H_2O \rightarrow N_2 + 4H^+ + 4Cl^- + C_xO \tag{9.11}$$

where C_x denotes the surface carbon atoms. The rate of reactions (9.10) and (9.11) is controlled by such factors as the properties of the active carbon surface (its chemical and capillary structure), the temperature of the process, the pH of the treated water and the presence in water of impurities other than chlorine that are adsorbed on active carbon. The optimization of the process should, therefore, be always preceded by preliminary investigations.

One should expect that in view of the continuously increasing contamination of water drawn and conditioned for municipal purposes, the application of active carbon in both existing and planned plants will be more and more frequent. In the USA, for example, the quantity of active carbon utilized in water and waste water treatment processes amounts to one fourth of the entire output of all types of active carbon.

Utilization of adsorption on active carbon in waste water treatment and water technologies. The rapid development of industry together with its large concentration in certain areas brings serious hazards to the natural environment. Pollution with liquid wastes of natural waters constituting the source of potable water for most large cities is a particularly serious problem. First of all this concerns large urban agglomerations of highly industrialized countries where, on one hand, the demand for municipal water is very large and, on the other, there is a heavy concentration of industry. Liquid wastes from this industry together with municipal sewage constitute a huge amount of heavily polluted water, and the effective purification of this water becomes problematic. In most cases biological and chemical methods of water treatment are insufficient to enable their safe disposal to water basins or the use of the industrial water in closed cycles. Among the established methods of water regeneration, adsorption on active carbon has found major application in the removal of harmful substances (most often detergents, dyes, aromatic hydrocarbons, esters and pesticides as well as heavy metal ions). It usually constitutes one of the several steps of water processing in waste-water treatment plants. Adsorption on active carbon is most commonly applied in the following technological systems of waste-water treatment:

(a) In the third step of mechanical-biological waste-water treatment plant, in two main versions: directly after biological treatment and filtration or after mechanical-biological treatment, coagulation and filtration (Fig. 9.3).

(b) As the final process of physico-chemical treatment of industrial liquid wastes or municipal sewage of such high concentrations of toxic substances that biological treatment is rendered impossible (Fig. 9.4).

Sec. 9.2] **Adsorption from liquid phase** 241

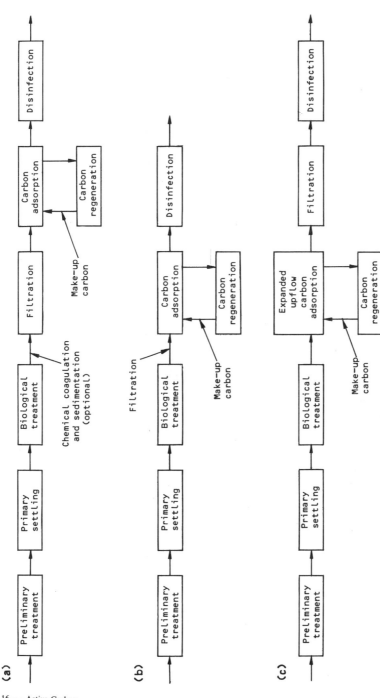

Fig. 9.3 Points at which active carbon is applied in different technological systems (a–c) of mechanical and biological water treatment (after Culp and Culp [22]).

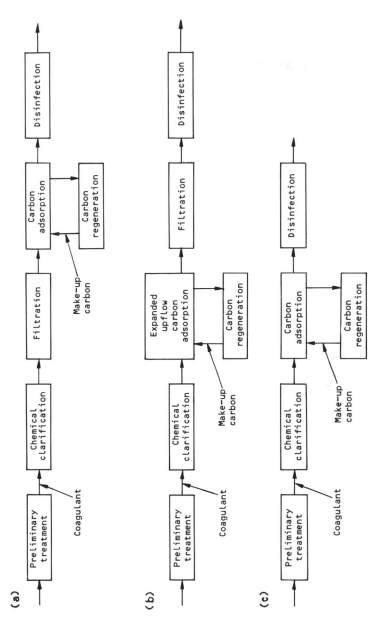

Fig. 9.4 Points at which active carbon is applied in different technological systems (a–c) of physico-chemical water treatment plants (after the U. S. Environmental Protection Agency [23]).

(c) As a process supporting biological treatment in which active carbon is added to the chamber containing activated sludge, and biological decomposition of organic compounds is accompanied by adsorption.

In the systems described it is possible to apply both powdered and granulated active carbon.

Adsorption on powdered active carbon is carried out in separate reactors often coupled with sedimentation tanks (as in one third of the systems described) or simultaneously with treatment on activated sludge, or possibly together with the precipitation and coagulation processes. The contact time of waste waters with active carbon range from 10 to 30 min and the carbon doses are 100–1000 g m^{-3}. After sedimentation of the carbon, filtration is still necessary. The main advantage of using powdered carbon are the significant flexibility of the system, enabling fast modification of the carbon dose depending on the composition of the waste water and the low investment costs. The major disadvantages are the impracticability of carbon regeneration, the high costs of the carbon, operational difficulties and problems with the removal of the smallest carbon particles from the water. Because of these disadvantages, powdered carbon has not yet found broader application in highly efficient technologies of waste-water treatment. It is, however, often used in water conditioning stations particularly for periodical, incidental adsorption of impurities.

Granulated active carbon does not exhibit the disadvantages characteristic of powdered carbon. In systematic use it is many times cheaper, so it is to be expected that it will become in the future the main adsorbent used in highly effective technologies of liquid waste treatment and water regeneration [24].

Application of granulated active carbon requires that the contact time of the liquid waste and carbon be provided for when designing the columns (contactors). It is assumed that the value of this parameter, related to the average flow rate, should be at least double that of the experimental minimum contact time which depends primarily on the adsorbability of the impurities to be removed and on the structure and specific surface area of the carbon. The height of the adsorption bed in the column is usually 3 to 10 m and the number of steps in the adsorption process 1 to 2 (in physicochemical treatment plants, sometimes 3). The major remaining process parameters involving the use of granulated carbon are:

(a) The hydraulic load dependent on the mode of bed operation.

(b) The mean diameter of the carbon granules (usually ranging from 0.5 to 2.5 mm).

(c) The degree of bed expansion (usually 15 to 30%).

(d) The column washing intensity (the amount of clean water passed through the column in a period of 10–15 min does not usually exceed 5% of total production of the plant).

(e) The frequency of washing (washing is applied every second day or every few days).

The waste-water treatment plants, both existing and under design and construction, with active carbon filling operate according to one of the following technological systems:

(a) Pressure systems with the flow of the liquid waste either upwards or downwards.

(b) Gravitational contactors with the flow of liquid waste either upwards or downwards.

(c) Columns with an upward flow of the liquid waste and with a compact or expanded bed.

(d) Columns coupled either in-parallel or in-series (single or multi-step).

The contact columns filled with active carbon may be either made of steel or concrete. Depending on the hydraulic load, the columns with an upward flow of waste water are designed in three possible versions: with a compact bed for small loads, with an expanded bed for medium-size loads, and with a mobile bed resting on the roof for large hydraulic loads. However, columns with an expanded bed do not retain suspensions, so in this case filtration must follow adsorption. Adsorption in columns with a downward flow is usually preceded by filtration. In systems where the filtration step is omitted, the removal of suspensions is conducted in the adsorption columns which, however, increases the rinsing frequency. Exchange of the spent carbon in such systems requires their temporary shut-down. An optimal utilization of the carbon is achieved in a system of two columns coupled in series.

The main advantage of technological systems with an upward flow of waste water as compared with those with downward flow is their high effectiveness of carbon utilization. This is due to the fact that in this case the process runs in a counter-current system, with continuous feeding of fresh carbon from the top of the column and removing spent carbon from the bottom without interrupting normal operation. In the counter-current technique the consumption of carbon is significantly reduced.

Comparison of the gravitational and pressure systems leads us, on the other hand, to the following conclusions. The gravitational system is cheaper to operate. The columns may be made of concrete which is more economical. However, to minimize the hydraulic losses, more effective removal of suspensions is necessary. Pressure columns, on the other hand, give more flexibility and enable work when pressure losses are higher. Regeneration of the spent carbon and its recycling to the column with simultaneous compensation of the losses is usually applied in all the systems discussed [2].

The global number of highly efficient water treatment plants (both industrial and municipal) utilizing adsorption on active carbon is growing continuously each year. This technology was first introduced in highly industrialized countries where the problem of shortage of natural water resources is particularly critical.

9.2.3 Chemical and Pharmaceutical Industries

Decolorization of organic substances is one of the oldest and most widespread applications of adsorption on active carbon. Removal of polymeric impurities (high molecular mass resins) producing dark colours and bringing about serious technological problems, particularly during crystallization, is a good example. Apart from removal of colour, adsorption on active carbon also enables clarification of a solution due to elimination of colloidal substances. Such a process is often conducted in a batch mode. A certain amount of powdered active carbon is added to a liquid to be purified and after sufficient time the spent carbon is separated on a filter press or centrifuge. The process is usually conducted at room temperature since higher temperatures reduce the decolorization efficiency. Heating is applied only in the case of high viscosity liquids to improve the conditions for diffusion (only when dilution in organic solvents is impossible). The efficiency of the decolorization process also depends on such factors as the type of solvent and the pH of the solution. The content of mineral impurities (ash) in active carbon and its chemical composition are also important because of the danger that the product being purified may become contaminated with salts extracted from the carbon. In some cases active carbon may cause catalytic oxidation of certain compounds, which may lead to a partial decomposition of the product and in consequence to a lower yield. Reducing agents are added in these cases to prevent such losses.

The pharmaceutical industry is particularly demanding as regards the application of active carbon adsorption processes. This involves, among other problems, the accuracy of dosing the adsorbent. The amount of carbon added should not be larger than needed for efficient decolorization, as otherwise a part of the product may also be adsorbed. In each case the dose should be optimized on the base of laboratory tests in which the amount of impurities to be adsorbed has been determined. The following compounds are good examples of the use of active carbon in decolorization processes: glycerol, betaine, lactic acid and its salts, glutamic acid and tartaric acid.

Another area of application of active carbon is the recovery (separation) of compounds from dilute solutions. One should note, however, that although used in the laboratory, it is very difficult to scale-up this process to the industrial scale. The adsorbed substances are strongly bound to the carbon surface, and hence their recovery involves significant losses and consequently the overall process is rather expensive. Another application of adsorption, namely chromatographic separation of such substances as alkaloids, vitamins or vegetable pigments is less effective than extraction with suitable solvents. The same can be said of antibiotics. Initially both penicillin and streptomycin were separated by adsorption on active carbon and removed by extraction with a suitable solvent. In more advanced technologies these processes are replaced by extraction. Active carbon is used only for removal of undesired impurities from penicillin which has already been extracted. In general, adsorption on active carbon is rarely utilized in the production of antibiotics. The separation of closely structurally-related antibiotics is an exception. More detailed information concerning the application of active carbon in the pharmaceutical industry as well as in several other branches of the chemical industry can be found in the literature [4, 25].

9.2.4 Medicine

Active carbon as an adsorbent was probably first applied in medicine. The students of Hipocrates recommended the dusting of wounds with charcoal as a means to remove their unpleasant smell. Systematic research on the antitoxic properties of active carbon began at the beginning of the twentieth century. Adsorption on active carbon of such toxic substances as heavy metal salts, alkaloids, barbiturates, phenols and alcohols as well as all sorts of insecticides and defoliants has been considered [26–32]. The real career of active carbon in medicine started in the 1950's when Arwall proposed the purification of the blood of poisoned patients by passing it through a column filled with active carbon [33]. This concerned particularly those cases when oral application of carbon was ineffective. This concept was supported and developed by the Greek scientist Yatzidis [34–36] who, together with his coworkers, succeeded in saving the lives of two patients poisoned with barbiturates by this method. The authors describe a 600 cm^3 cylindrical column of their design filled with 200 g of active carbon of a particle size 0.4 to 1.4 mm, and additionally equipped with two filters. The column was coupled to a pump and its inlet and outlet were connected to the patient's artery and vein, respectively. Yatzidis called his apparatus a 'carbon kidney' [35, 36]. He postulated that active carbon may be used for the removal from blood not only of externally introduced poisons but also of toxins which appear in the blood due to the failure of the patient's kidneys. Results of other workers [37–39] show the prospects of applying the 'carbon kidney' in clinical practice, but they also point to the disadvantages of this method. Together with noxious substances the 'carbon kidney' retains certain vital components of blood. Some blood clots were trapped between the carbon pellets and this fraction was irretrievably lost for the patient. These authors also show [38] that, in rabbits treated with barbiturates, the poison is indeed removed from the blood, but carbon powder was found in the animals' liver, lungs, spleen, brain and kidneys. Further research has shown that in every type of granulated active carbon a certain amount of powdered material occurs which cannot be removed by any amount of washing [40]. This powder is formed as a result of attrition of the pellets during washing, during passing blood through the bed, and especially during transport. The presence of powder in every brand of active carbon is the main obstacle to its widespread application as a means of purification of the blood of a living organism. As mentioned above the carbon powder accumulates in various organs, inducing serious problems. Therefore, methods of carbon treatment were sought in which the formation of powder from the pellets would be prevented and the process of blood clotting eliminated. For instance, it was proposed to place semipermeable membranes on the pellet surface [41, 42]. The thickness of such membranes was less than 50 nm, which ensured a sufficient rate of diffusion of adsorbed metabolites. Additionally pellets were covered with polymer membranes markedly improving the miscibility of blood with carbon. Treatment of pellets with active substances which gave the pellets a negative charge brought further improvement. The deposition on the carbon granules of natural polyampholyte-albumins is also advantageous. At the beginning of the seventies Chang proposed an original blood purifying apparatus utilizing carbon granules coated in this way [43]. This apparatus was successfully used in the therapy of acute poisoning and chronic kidney failure. This author showed that columns filled with 30 g of active carbon pellets encapsulated in albumin are more efficient than the currently used 'artificial kindey'. However, the coating of pellets with albumin is expensive and the resulting adsorbent is not very stable.

In an earlier study [44] it was shown that the intensity of blood clotting is strongly related to the smoothness of the pellet surface which, incidentally, also influences powder formation. The electrical properties of blood itself are also important. They may be modified by applying pharmacological means which increase the suspensive stability of blood and improve its circulation in blood vessels.

Nikolaev et al. [45] presented in 1976 a list of adsorbent properties which have to be determined before any adsorbent may be used for human blood purification. These are:
1. The physical and mechanical properties of the adsorbent granules
   (a) their texture,
   (b) shape, and
   (c) attrition strength.
2. The physicochemical properties of the adsorbent
   (a) its specific surface area and distribution of pores as a function of their effective radii,
   (b) the chemical structure of the carbon surface (degree of oxidation, character of the surface functional groups),
   (c) the amount and composition of impurities,
   (d) the standard kinetic adsorption characteristics.
3. Interactions of blood with the adsorbent and other biological characteristics
   (a) the dynamic resistance of the column and the mass increase of the adsorbent after adsorption,
   (b) the dynamic properties of blood proteins,
   (c) the charge of adsorbent granules in the biological liquid,
   (d) the ability of the adsorbent to adsorb the specific metabolites from plasma and its substitutes,
   (e) the occurrence of biological side-effects controlled by non-specific adsorption,
   (f) toxicity of the adsorbent due to the presence of impurities.

The criteria presented allow us to judge the suitability of a carbon adsorbent for the process of removing toxic substances, both exogenous, and endogenous, from blood. A list of some exogenous and endogenous

Table 9.3

Substances removed from blood or lymph following disease and poisoning (after Buruskina [46])

Substances removed	Liver failure	Kidney failure	Exogenous poisoning	Radiation burns in oncology	General disturbances in acid-base equilibrium of salt exchange	Stored blood
Ions K^+, Cl^-, $H_2PO_4^-$, HCO_3^-				+	+	+
Nitrogen-containing compounds:						
ammonia		+		+		+
creatine	+	+	+	+		
urea	+	+	+	+		
uric acid	+	+	+	+		+
Organic acids:						
lactic				+	+	+
pyruvic				+	+	+
acetoacetic					+	
β-hydroxybutyric					+	
aminoacids	+			+		
Proteinaceous substances:						
oligomers of nucleic acids				+		
fragments of cell membranes	+			+		+
tissue protein				+		
free haemoglobin				+		+
bilirubin	+		+	+		
Exogenous phenols	+		+	+		
Cytostats	+			+		

toxic substances that sometimes have to be removed simultaneously from blood is given in Table 9.3 [46]. In Table 9.4, on the other hand, the maximum adsorption on active carbon of certain low-molecular toxic substances are given in terms of their concentration dependence in blood [47].

It seems that so far the optimum properties, in terms of human blood purification, are exhibited by the adsorbent described by Strelko [48] and by Nikolaev and Strelko [49]. This adsorbent, produced from

Table 9.4

Adsorptive capacity of active carbon (after Walker et al. [47])

Compound	Blood concentration mg dm^{-3}	Adsorption mg (200 g)$^{-1}$
Creatine	110–250	600–1500
Uric acid	50–150	400–1200
Indican	4–5	25–30
Phenols	30–40	150–175
Guanidine	30–40	85–110
Organic acids	15–20	30–35
Urea	7–4	2–3
Barbiturates	75–300	500–2000
Salicylates	250–750	2000–3000

a styrene-divinylbenzene copolymer, is in the form of spherical granules with an exceptionally smooth and hard surface which are not coated with any film whatsoever. Since this carbon is not subjected to any additional treatment, the normal side-effects do not occur, and moreover the adsorbent is cheaper compared with others produced for the same purpose. Today this brand of carbon is produced in the Soviet Union on a commercial scale.

9.3 ACTIVE CARBON AS AN ELECTRODE MATERIAL

In view of their relatively high electrical conductivity, well-developed surface area, good catalytic properties, high stability in contact with acids and bases, as well as of their ready availability and relatively low price, active carbon and carbon black find widespread application as electron transfer materials in chemical devices (recently especially in fuel cells). They are used chiefly in the manufacture of porous gas electrodes, mainly oxygen electrodes. In such electrodes the electrode process proceeds at the interface of three phases: solid (carbon material on which a catalyst may additionally be deposited), liquid (electrolyte solution) and gaseous (oxygen and air). The system considered is very complex and the mechanism of some of the processes involved is not yet fully understood. The successful manufacture of carbon diffusion oxygen electrodes characterized by good and reproducible parameters for operation in a cell, requires knowledge of the kinetics and mechanism of the electrode processes, the transport macrokinetics, the dependence of the electrode performance on the carbon material, and last but not least, the design and manufacturing technology of the electrode. The wide practical application of fuel cells, in which the diffusion oxygen electrode is one of the most important and most troublesome elements, strongly depends on a good understanding and solution of these problems. The most advanced research today, both theoretical and technological, is concerned with the application of the carbon electrode in small zinc-air or hydrogen-oxygen (air) cells filled with a basic electrolyte. The suitability of oxygen diffusion electrodes in fuel cells in which methanol is the fuel (other alcohols and hydrocarbons may also be used) is also being investigated. In all these types of cell, oxygen is supplied to the electrode by diffusion from air. In batteries composed of many cells air is usually blown by a fan. The diffusion carbon electrodes used in fuel cells show high durability. Idle carbon electrodes may remain in contact with aggressive electrolyte solutions without harm for relatively long periods. Their longevity during stable operation under current load ranges from one to two years.

Apart from their application in fuel cells, diffusion oxygen electrodes made of carbon may also be used as indicator electrodes in the electrochemical sensors of oxygen analysers. Since the early sixties a constant development has taken place of various electrochemical methods for the quantitative analysis of oxygen. Several designs differing as regards construction, principle of operation, electrode material and electrolyte solution have been proposed since [50].

The possibility of the practical application of active carbon (or carbon black) as an electrode material is one of the major reasons for the large scientific interest in electrochemical reduction of oxygen (the most readily available and cheapest oxidant) on the surface of carbon materials. The application in the recent years of such research methods as the rotating disc electrode, the rotating ring disc electrode, single-sweep voltammetry or isotopic labelling with ^{18}O clarified the mechanism of the electrochemical reduction of oxygen [51]. This is a chain reaction in which the following steps can be distinguished:

(i) diffusion of oxygen to the carbon surface when oxygen is adsorbed according to the equation:

$$C_x + O_2 \rightarrow C_x - O_{2(ads)} \tag{9.12}$$

(where C_x is the active site on the carbon surface),

(ii) electron transfer from the carbon electron gas to the adsorbed molecules of oxygen

$$C_x - O_{2(ads)} + e^- \rightarrow C_x - O_{2(ads)}^- \tag{9.13}$$

and,

(iii) the reaction of the adsorbed O_2^- ions with water molecules according to the equations:

$$C_x-O_{2(ads)}^- + H_2O + e^- \rightarrow HO_2^- + OH^- + C_x \quad (9.14)$$

or

$$C_x-O_{2(ads)}^- + H_2O + e^- \rightarrow C_x-O_{(ads)} + 2OH^- \quad (9.15)$$

Assuming interaction between the ionized molecules of oxygen adsorbed on the surface, the last step of the process can be represented in the following scheme:

$$\begin{array}{ccc} C_x-O_{2(ads)}^- & & C_x+HO_2^- \\ \cdot & & \cdot \\ \cdot & +H_2O \rightarrow & \cdot +OH^- \\ \cdot & & \cdot \\ C_x-O_{2(ads)}^- & & C_x+O_2 \end{array} \quad (9.16)$$

One of the products of reactions (9.14) and (9.16) is peroxide ions. Their reactivity has a significant impact on the current-voltage characteristics of the carbon oxygen electrode. On the other hand, hydrogen peroxide formed in this reaction brings about slow oxidation of carbon, makes it more hydrophilic and in effect reduces the duration of its stable operation. It is important therefore to remove H_2O_2 from the reaction medium.

The decomposition of hydrogen peroxide may proceed via different mechanisms. Its catalytic decomposition on active carbon proceeds in a basic environment, but if a catalyst is deposited on the active carbon, then it also decomposes in acidic or neutral media. The reaction, which is regarded as a first-order process with respect to the hydrogen peroxide activity, may be presented schematically as follows:

$$C_x + H_2O_2 \xrightarrow{slow} C_x \ldots O + H_2O \quad (9.17)$$

$$C_x \ldots O + H_2O_2 \xrightarrow{fast} C_x + O_2 + H_2O \quad (9.18)$$

where $C_x \ldots O$ denotes an oxygen atom chemisorbed on carbon. One should note that the acidic functional groups present on the carbon surface inhibit the above reaction, whereas basic functional groups accelerate it. Extensive research on the effect of such factors as the capillary structure of the carbon, the chemical structure of its surface and the type of catalysts deposited on the electrochemical reduction of oxygen and H_2O_2 decomposition has resulted in the design of oxygen diffusion electrodes with very desirable current-voltage characteristics. Among the catalysts increasing the electrocatalytic activity of carbon oxygen electrodes used in fuel cells are: (i) in basic media: silver and its salts, spinel-type oxides, MnO_2, phthalocyanines and polyphthalocyanines of transition metals, and (ii) in acid media: platinum, phthalocyanines and polyphthalocyanines of transition metals. The mechanism of electrochemical reduction of oxygen on carbons impregnated with catalysts depends strongly on the type of carbon and on the activity and distribution of the catalyst.

As an example of the design of a diffusion oxygen electrode [52] we can consider the vessel with a carbon electrode used in the zinc-oxygen cell shown in Fig. 9.5. The catalysts deposited on three different active carbons (used together with teflon to make the inner layer of the diffusion oxygen electrode) tested at the research stage were: silver, cobalt phthalocyanine and cobalt octamethoxyphthalocyanine. It was found that each of the organometallic catalysts showed an electrochemical activity similar to that of silver. In concluding one should emphasise that although many of the problems associated with electrochemical reduction of oxygen on carbon electrodes have been successfully solved, the technology of manufacturing carbon diffusion electrodes still presents a significant barrier to their mass production. The main problem consists in obtaining satisfactory reproducibility of the properties of carbon materials (most often active carbon), optimal selection of the carrier and catalyst, and achieving good stability of the final product.

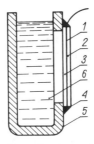

Fig. 9.5 Cell with oxygen electrode: 1 — nickel wire mesh, 2 — outer porous hydrophobic layer, 3 — inner layer consisting of Teflon and active carbon with deposited catalyst, 4 — adhesive-bonded joint, 5 — organic glass housing, 6 — aqueous KOH solution (concentration 7 mol dm^{-3} (after Tomassi et al. [52]).

9.4 UTILIZATION OF ION-EXCHANGE PROPERTIES OF ACTIVE CARBON

For some decades the ion-exchanges properties of active carbon have been the object of significant interest, both theoretical and practical [53–55]. Compared to many other ion-exchanging materials, both synthetic and mineral, active carbons exhibit a number of advantages. Their specific ion-exchange properties may be readily tailored in the desired direction and their ion-exchange capacity may also be controlled. Additionally, active carbon shows high chemical resistance to aggressive solutions. Active carbons (both commercial and prepared on a laboratory scale) exhibit some ion-exchange properties even when not subjected to any additional treatment. However, in order to enhance these properties, as well as to develop them in the desired direction, the carbon surface is usually chemically modified. This treatment consists in incorporating different heteroatoms into the surface with simultaneous chemical bonding of these atoms onto the carbon lattice. Most often these are oxygen, nitrogen, phosphorus or sulphur atoms which are usually built into the carbon structure in the form of typical organic functional groups. Carbons with oxygen-containing surface groups are of particular importance. Both untreated and chemically modified active carbons exhibiting ion-exchange properties currently find widespread practical applications in various technologies. The most important are removal of the trace amounts of heavy metals (toxic) from waste waters and during conditioning of potable water, purification of solutions of inorganic salts, recovery of small amounts of valuable metals in the chemical industry, and preconcentration of microadmixtures before applying typical instrumental analytical techniques. In order to discuss in more detail the application of active carbon in ion exchange we shall consider several selected examples of such applications. They concern cation-exchange properties of carbons with an oxidized surface.

Oxidation of carbons leading to a significant increase of the number of surface oxygen-containing functional groups of acidic character (exhibiting cation-exchange properties) is relatively easy from the technological point of view and does not require any complicated equipment. For technological and economic reasons oxidation with nitric acid, sodium hypochlorite, hydrogen peroxide or air is usually recommended. In this way a carbon of ion exchange capacity of 2–3 mmol g^{-1} is usually obtained. The use of each of these oxidants is associated, however, with certain disadvantages. Oxidation with nitric acid, for instance, gives a suitably rigorous surface oxidation but brings about at the same time a large concentration of humins on the oxidized carbon surface. These substances also occur in the cases of other oxidants but usually in much smaller amounts. The removal of these substances (which may contaminate the solutions during the use of the carbon as an ion exchanger) from carbon requires heat treatment at 300–350°C under vacuum (which may decrease the ion-exchange capacity of the material), or washing with a sodium hydroxide solution and followed by washing with aqueous hydrochloric acid. Oxidation with air, conducted at 400–500°C, causes a significant burn-off of the carbon (the weight loss may exceed 50%) due to the simultaneous generation of surface oxides and gaseous CO and CO_2. On the whole, however, the methods of oxidation described are relatively simple as compared with other ways of modifying the carbon surface. Largely due to this and to its strong cation-exchange properties, oxidized active carbon is finding ever-increasing practical application. The very high cation-exchange selectivity of these materials is also of great importance. The $Me^{z+} - H^+$ exchange constants with respect to particular cations may differ even by several orders of magnitude. The exchange constants for oxidized active carbons as determined by chromatographic and adsorption methods may be arranged in the order of increasing ion-exchange capacity as follows [51]:

$NH_4^+ < Na^+ < Rb^+ < Cs^+ < Mg^{2+} < Cd^{2+}, Mn^{2+} < Sr^{2+} < Ca^{2+}, Zn^{2+}, Fe^{2+} < Ni^{2+}, Al^{3+} < Y^{3+} \leqslant Cr^{3+} \leqslant Be^{2+} < Cu^{2+} < Fe^{3+}$.

We shall now consider several examples of applying oxidized carbon as an ion-exchanging material [53].

During the electrolytic production of chlorine and sodium hydroxide from brine, the removal of undesirable components (e.g. poisons such as Mo, V, Cr in the mercury-based method of electrolysis) is a significant problem. Noxious microimpurities which may disturb the technological process even at concentrations lower than 10^{-5} g dm^{-3} are successfully removed by oxidized active carbon acting as a highly selective cation exchanger. The spent carbon is usually regenerated by washing with hydrochloric acid, after which it may be used again in the process. Carbon ion-exchangers are applied not only in the purification of substrate (brine) but also in the removal of impurities from the product, i.e. solutions of sodium or potassium hydroxide. The superiority of active carbon over synthetic or mineral ion exchangers, which are much less resistant to concentrated solutions of strong bases, is revealed here in full. Active carbon ion exchangers are also suitable for removing from such solutions alkaline earths or rare earths, as well as Fe, Co, Mn and other metals. Purification of hydroxide solutions by means of active carbon requires, however, very careful removal of humins from its surface, otherwise these substances could migrate into the solutions being purified. Active carbon may also be used for purification of solutions of other hydroxides (e.g. of LiOH from Mg, Al, Ca impurities) or alkali metal carbonates. Products of high purity may be obtained in this way.

Another application of carbon ion-exchangers (oxidized carbons) is the preparation of compounds of very high purity for special uses. Synthetic ion exchangers cannot be used in these cases because of the danger of contamination of the solution with organic impurities (e.g. monomers). The following processes may serve as examples of extreme purification: the preparation of hydrogen phosphates of alkali metals which serve as raw materials for the production of metaphosphates used in the manufacture of special glass, the preparation of salts of rare earth metals which serve as raw materials for the production of piezoelectric and seignette-electric materials, the preparation of ammonium carbonate, potassium chloride and zinc sulphate necessary for the manufacture of cathodoluminophors, and of many other salts used as raw materials in various technologies.

Oxidized carbons can also be used in analytical chemistry as cation exchangers. They enable preconcentration of the substance to be analysed thus increasing considerably the sensitivity of the analysis. Note that thanks to their selectivity, carbon ion-exchangers are also very useful in cases when small amounts of additives are to be analysed under conditions of a strong background of the main component of the solution. Concentration of traces of cations by means of oxidized carbon, as a method supporting analytical techniques, finds widespread application in various chemical processes. It is applied in the manufacture and quality control of high-purity substances used in the microelectronics and nuclear industries, in the production of optical and luminescent materials, in the determination of microelements in water or soil, and in many other branches of inorganic technology. In all these applications carbon has to be very carefully purified from mineral matter (ash) by repeated washing with, e.g., hydrochloric and hydrofluoric acids. This is necessary because of the possibility of contaminating the solution of the substance to be analysed or purified with the soluble components of ash. In most cases the contact of carbon with the solution has a dynamic nature (the solution flows through a column with a stationary bed of the cation exchanger).

Incorporation of heteroatoms other than oxygen into the carbon surface enables the derivation of ion exchangers (both anion and cation) of strongly diversified properties. To incorporate nitrogen into its surface, carbon (oxidized or not) is treated with ammonia at high temperatures [55]. In this process various pyridine and piperidine type-structures, and, in the case of preoxidized carbon, also amide and imide groups, as well as ammonium salts of carboxylic acids are formed on the carbon surface. Carbon preparations containing nitrogen atoms incorporated into the carbon skeleton may also be obtained by using various polymers containing nitrogen in the molecule (such as polyacrylonitrile) as substrate. Active carbons containing nitrogen bonded to their surface exhibit anion-exchanging properties. Another heteroatom, i.e. sulphur, is also incorporated into the carbon surface at higher temperatures by using such sulphurizing agents as sulphur vapour, carbon disulphide, sulphur dioxide or hydrogen sulphide [56]. The sulphur atoms are bonded mainly at unsaturated sites of the surface but they may also form surface compounds such as sulphides, thiophenols, thioethers, thioquinones and sulphoxides. Sulphur-containing carbons are characterized by diversified ion-exchange properties depending on the amount of bonded sulphur and its mode of bonding.

The phosphorizing of active carbon with phosphoric acid or with phosphorus trichloride vapours at high temperatures results in the formation on the carbon surface of complex oxygen-phosphorus structures which give the carbon preparations cation-exchange properties. Impregnation of carbon with different chemical compounds (usually organic) also creates great possibilities for modifying the carbon properties.

It is noteworthy that to date possibilities of applying active carbon as an ion-exchanger have not yet been fully explored. Carbon ion-exchangers constitute a valuable supplement to mineral and synthetic materials, and in some areas have proved to be even better due to their individual properties.

References

[1] Keltsev, N. V., *Podstawy techniki adsorpcyjnej* (Principles of Adsorption Technology) WNT, Warsaw 1980.
[2] Kienle (von), H., Bäder, E., *Aktivkohle und ihre industrielle Anwendung*, Ferdinand Enke Verlag, Stuttgart 1980.
[3] Jüntgen, H., *Carbon* **15**, 273 (1977).
[4] Smíšek, M., Černý, S., *Active Carbon*, Elsevier, Amsterdam-London-New York 1970.
[5] Knoblauch, H., Reichenberger, J., Jüngten, H., *Gas Erdgas* **116**, 382 (1975); *Erdöl und Kohle* **28**, 426 (1975).
[6] Bałys, M., Czapliński, A., Ziętkiewicz, J., *Przem. Chem.* **62**, 516 (1983).
[7] Jankowska, H., Choma, J., *Biul. WAT* **31**, No. 6, 109 (1982).
[8] USA Patent 1 519 470, 1924.
[9] FRG Patent 1 087 579, 1961.
[10] Choma, J., Horak, J., Rozmarynowicz, M., *Biul. WAT* **32**, No. 8, 75 (1983).
[11] Choma, J., Doctoral Thesis, Technical Military Academy, Warsaw 1981.
[12] Zabar, J. W., *Mechanism of Chemical Removal of Gases* **7**, 161 (1973).
[13] Jonas, L. A., Rehrman, J. A., *Carbon* **10**, 657 (1975).
[14] Barnir, Z., Aharoni, Ch., *ibid.* **13**, 363 (1972).
[15] Chiou, C. T., Reucroft, P. J., *ibid.* **15**, 49 (1977).
[16] Lew, R. B., *Int. Sugar. J.* **85**, 323 (1983).
[17] Anderson, A.S.G., *Refining of Oils and Fats*, Pergamon Press, London 1953.
[18] Singleton, V. L., Draper, D. E., *Am. J. End. Viticult.* **13**, 114 (1962).
[19] Connick, R. E., Yuan-Tsan Chia, *J. Am. Chem. Soc.* **81**, 1280 (1959).
[20] Suidan, M. T., Snoeyink, V. L., Schmitz, R. A., *Envir. Eng. Div.* **4**, 103 (1977).
[21] Kim, B. R., Snoeyink, V. L., Schmitz, R. A., *J. Water Poll.* **1**, 50 (1978).
[22] Culp, R. L., Culp, G. L., *Advanced Wastewater Treatment*, Litton Educational Publishing, New York 1971.
[23] Process Design Manual for Carbon Adsorption, Technology Transfer, U. S. Environmental Protection Agency (Oct. 1973).
[24] Kurbiel, J., *Węgiel aktywny. Problemy badawcze i wdrożeniowe w gospodarce wodnej* (Active Carbon. Research and Implementation in Water Technology), Proceedings of Symposium, Zakopane 1980, p. 160.
[25] Hassler, J. W., *Activated Carbon*, Chemical Publishing Co., New York 1963.
[26] Lichwitz, L., *Therap. d. Gegenw.* 543 (1908).
[27] Joachimogly, G., *Biochem. Z.* **77**, 1 (1916).
[28] Dingemanse, E., *Am. J. Physiol.* **90**, 329 (1929).
[29] Andersen, A. H., *Acta pharm. et tox.* **4**, 275 (1948).
[30] Chin, L., Picchioni, A. L., Duplisse, B. R., *Fd. Proc.* **26**, 761 (1967).
[31] Decker, W. J., Combe, H. F., Corby, D., *Tox. a. Appl. Pharm.* **13**, 454 (1968).
[32] Chin, L., Picchioni, A. L., Duplisse, B. R., *ibid.* 16, 786 (1970).
[33] Alwall, N., *Acta Med. Scand.* **180**, 593 (1952).
[34] Yatzidis, H. et al., *Lancet* **2**, 216 (1965).
[35] Yatzidis, H., *Nephron* **1**, 310 (1964).
[36] Yatzidis, H., *Proc. Europ. Dial. Trans. Ass.* **1**, 83 (1964).
[37] Dunca, G., Kolff, W. J., *Trans. Amer. Soc. Artif. Intern. Organs* **11**, 179 (1965).
[38] Hagstam, K. E., Larsson, L. E., Thysell, H., *Acta Med. Scand.* 180, 593 (1966).

[39] Andrade, J. D., Kunitomo, K., Van Wagenen, R., et. al., Trans. Amer. Soc. Artif. Intern. Organs **17**, 222 (1971).
[40] Andrade, J. D., Kunitomo, K., Van Wagennen, R., et al., ibid. **18**, 473 (1972).
[41] Levine, S. N., La Course, W. C., J. Biomed. Mater. Res. **1**, 275 (1967).
[42] Sparks, R. E., et. al., Trans. Amer. Soc. Artif. Intern. Organs **15**, 353 (1969).
[43] Chang, T. M. S., Artificial Cells, Thomas, Springfield, Ill., 1972.
[44] Kolff, W. J., Circulation **17**, 702 (1958).
[45] Nokolaev, V. G., Strazhesko, D. N., Strelko, V. V., et al., Adsorbtsiya i adsorbenty **4**, 24 (1976).
[46] Buruskina, T. N., ibid. **9**, 77 (1981).
[47] Walker, J. M., Denti, E., Van Wagenen, R. D., et al., Kidney Inter. Suppl. **7**, 320 (1976).
[48] Strelko, V. V., Galinskaya, V. I., Davydov, V. I., et al., Adsorbtsiya i adsorbenty **4**, 29 (1976).
[49] Nikolaev, V. G., Strelko, V. V., Gemosorbtsiya na aktivnykh uglakh (Chemisorption on Active Carbons), Naukova Dumka, Kiev 1979.
[50] Zwierzchowska-Nowakowska, Z., Jankowska, H., Przem. Chem. **56**, 79 (1977).
[51] Jankowska, H., Neffe, S., ibid. **59**, 591 (1980).
[52] Tomassi, W., Dąbrowski, R., Choroś, D., Roczniki Chem. **48**, 683 (1974).
[53] Tarkovskaya, I. A., Okislennyi ugol' (Oxidized Carbon), Naukova Dumka, Kiev 1981.
[54] Cheremisinoff, P. N., Ellerbush, F., Carbon Adsorption Handbook, Ann Arbor Science Publishers, Ann Arbor, Mich. 1978.
[55] Jankowska, H., Starostin, L., Przem. Chem. **62**, 440 (1983).
[56] Jankowska, H., Gajewski, M., Świątkowski, A., et al., Adsorptsiya i adsorbenty **7**, 10, (1979).

CHAPTER 10

Regeneration of Spent Active Carbon

10.1 THEORETICAL BASIS OF THE REGENERATION AND REACTIVATION PROCESS

Active carbons have been applied as adsorbents for hundreds of years but the regeneration of spent carbons was introduced 150 years ago.

The regeneration of the adsorbent consists of removing the adsorbed substances from its surface and in restoring, as far as possible, its initial adsorptive properties. In industrial practice we are concerned either with the recovery of any valuable materials adsorbed on the carbon surface or to use the same adsorbent many times for removing noxious substances. The regenerated adsorbent should therefore fully recover its initial adsorptive capacity or very nearly so.

The difficulty in solving this problem depends on whether we are dealing with physical or chemical sorption. Physical adsorption is a reversible process (although not in the thermodynamic meaning) and the removal of the adsorbate from the surface presents in this case no severe problem and may be realised, for instance, simply by heating. In contrast, regeneration in the case of chemisorption is a much more complex problem, due to the much stronger forces bonding the adsorbate molecules to the adsorbent surface. In the case of the physical process the attachment of the adsorbate is due to van der Vaals forces, hydrogen bonds, etc., whereas in chemisorption we are dealing with ionic or covalent bonding.

Klein [1] summarized this situation in the diagram shown in Fig. 10.1. Here Q_A is the heat of adsorption and ΔE_A is the activation energy of the adsorption process. In the case of physical adsorption, ΔE_A is small compared with the activation energy of chemisorption ΔE_C, and therefore the adsorbate can be removed fairly easily from the adsorbent surface, e.g., by heating, lowering the pressure, washing with solvent, etc.

Sec. 10.1] **Theoretical basis** 261

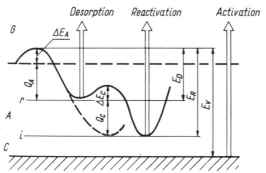

Fig. 10.1 Illustration of the concepts involved in regeneration of adsorbents: r — reversibly adsorbed adsorbate, i — irreversibly adsorbed adsorbate, G — gas phase, A — adsorbate, C — adsorbent (after Klein [1]).

The desorption energy is denoted in Fig. 10.1 by E_D which is equal to the sum of Q_A and the activation energy ΔE_A ($E_D = Q_A + \Delta E_A$). When the adsorbate molecules are weakly bonded to the adsorbent surface (physical adsorption) it is sufficient to lower the partial pressure of the adsorbate in the gas phase and hence shift the adsorption equilibrum towards desorption simply by lowering the overall pressure. If these energies are somewhat higher, desorption is effected by heating. This, however, may sometimes lead to the destruction of the adsorbate molecules and such a possibility should be taken into account. In the case of chemisorption the reversibility between the adsorption and desorption processes is incomplete. Supply of energy equal to the desorption energy does not now lead to complete regeneration of the adsorbent surface. Additional energy should be provided, denoted in Fig. 10.1 by Q_C, which represents the heat of chemical reaction of the adsorbate molecule with the adsorbent surface, i.e. Q_C is the heat of chemisorption. The process of removing the irreversibly bound adsorbate from the surface, a process designated reactivation by Klein, requires an expenditure of energy, denoted in Fig. 10.1 as E_R, which is equal to the sum of the energy of desorption and the heat of reaction ($E_R = E_D + Q_C$). We refer to reactivation when, in the processes of adsorption or desorption, chemical reactions take place and compounds of low volatility are formed that cannot be removed from the surface by conventional regeneration. Involatile compounds produce a significant lowering of the efficiency of active carbon, and their removal requires processes similar to those of activation, hence their name — reactivation processes.

Reactivation of the adsorbent is particularly important in the field of environmental protection since the water and air requiring purification before being released often contain substances which are very strongly bonded to the active carbon during adsorption and desorb with difficulty or not at all. Today there are possibilities of reactivating not only granular but also powdered active carbons in special processes and devices of advanced design. In many cases the boundary between regeneration and reactivation is very diffuse.

Complete desorption of substances from the active carbon surface proceeds with great difficulty. This is primarily due to the fact that the heat of adsorption and hence the heat of desorption is the greatest for small degrees of coverage of the adsorbent surface. In the words, the degree of desorption increases with temperature. However, since heating of the adsorbent layer to a high temperature is often undesirable, there always remains in the active carbon a certain amount of adsorbate which reduces in successive cycles the adsorptive capacity of the carbon. Apart from normal reversible adsorption controlled by the adsorptive properties of the carbon, irreversible adsorption occurs consisting of the accumulation in the pores of involatile products of polymerization and polycondensation of the adsorbate. These products are difficult to remove even if the active carbon is heated to 900°C. Usually good results are obtained by the action of steam heated to 700–900°C on the carbon. However, in all cases of regeneration and reactivation those process conditions should be established which exclude significant destruction of the carbon structure.

The conditions of regeneration and reactivation depend both on the kind of active carbon involved and on the kind of substance to be removed. For example, carbons activated with steam at 800–1000°C are more weakly reactive than active carbons with irreversibly adsorbed organic materials. Under appropriate conditions, reactivation with steam of the latter leads to decomposition of these materials; the involatile products first undergo carbonization, and then part of the generated elementary carbon is expelled in the form of gases produced in the reactions [2]:

$$C + O_2 \rightarrow CO_2 \tag{10.1}$$

$$C + CO_2 \rightarrow 2CO \tag{10.2}$$

Sometimes the activity of the active carbons is actually increased by the reactivation process. This is explained by the development of additional porosity of the active carbon due to the activation of the adsorbed organic substances.

When the initial raw material used for the production of some active carbons is the same, or when they were manufactured by the same method, their liability to reactivation is proportional to the specific surface area of the adsorbent as determined, e.g., by the BET method. Practice also shows that the rate of reactivation of macroporous active carbons is greater than that of microporous carbons.

Reactivation can be conducted in two ways: the first consists of reacting the irreversibly bonded adsorbate with the reactivating gas, and the second in a chemical reaction of the adsorbate with the reactivating liquid, the latter also involving extraction.

10.2 CONDITIONS OF THE REGENERATION AND REACTIVATION PROCESSES

The optimal conditions of the regeneration or reactivation process ensuring minimal losses of the carbon are determined in every case experimentally, three factors playing here a major role:

(1) the temperature of the oven,
(2) the time of regeneration or reactivation,
(3) the degree of saturation of the carbon with the adsorbed substance.

The first two are easy to check, control and optimize. It should be noted, however, that the choice of a given reactivation temperature, especially in large ovens, often leads to significant extension of the duration of the process.

Active carbons obtained from coal require greater regeneration (reactivation) energies than those obtained from coconut shells or wood. These differences affect the degree of burn-off of the active carbons depending on the reactivation temperature. Active carbons based on coal show lowest burn-offs. Brown coal-based active carbons show a small burn-off up to 800°C, but as the temperature is raised their burn-off increases significantly, achieving a dozen or so per cent at 900°C. Wood charcoal is still less resistant to temperature, and its burn-off may exceed 20% at 900°C.

The iodine numbers, characterizing the specific surface area, determined for carbons regenerated at various temperatures, show a distinct increase of the brown coal-based carbon burn-off with regeneration temperature, however, the greatest burn-off is observed for

Table 10.1

Iodine numbers of active carbons regenerated at various temperatures (after Kienle and Bäder [2])

Iodine number	Active carbon		
	coal-based	brown coal-based	charcoal-based
Initial value	1150	1170	1020
Value after heating at:			
700°C	1080	1270	1150
800°C	1090	1290	1160
900°C	1170	1330	1270
% maximum increase	1.7	12.0	24.5

charcoal (Table 10.1). The dependence of the weight-loss of carbon on the calcination temperature shows that the initial burn-off rate is greater than the final. For instance, reactivation of active carbon from charcoal at 840°C with a reactivating gas containing 40% of steam in a fluidized bed process produces after 10 min, 20 min, and 30 min losses of *ca.* 6 wt.%, over 8 wt.% and *ca.* 10 wt.% of carbon, respectively.

When evaluating active carbons we should consider not only their degree of burn-off during regeneration but also their adsorptive properties after reactivation since the adsorptive volume of a reactivated carbon is often greater when the burn-off is significant than for an insignificant burn-off. The product of the quantity of carbon obtained after reactivation (in. wt.%) and its adsorptive activity with respect to the initial carbon (in %) is a convenient parameter characterizing the general reactivation properties of a carbon. For example, this activity for brown coal-based active carbon is 12% in terms of the iodine number.

To estimate the adsorptive properties of the regenerated or reactivated carbon, especially when intended for water treatment, molasses, phenol, sodium dodecylbenzenesulphonate, antipyrine or tannin are recommended as adsorbates. For testing a granular active carbons, a precisely-determined amount of solution of a standard substance of known concentration is passed through the carbon layer of known height prepared from a weighed-out portion of the active carbon, and the final concentration of that substance is determined in the filtrate. The testing of powdered active carbons consists of adding an accurately-determined weighed-out portion of the carbon to a solution of a standard substance and in determining its final concentration after a given time.

The choice of test method depends on the particular application of the carbon. For example, in the case of active carbons used for purification of waste waters, one determines their total content of carbon originating from organic compounds, the total nitrogen content, the chemical demand for oxygen, the biological demand for oxygen, and the absorption of UV light at 220 and 278 nm.

It should be noted that active carbons in which macropores dominate regenerate more easily than those with fine pores, i.e. for comparable conditions and equal degrees of filling of their adsorptive volume. Most inorganic additives exert a catalytic effect on the regeneration or reactivation process, so sometimes the regeneration can be conducted under conditions milder than those expected from the degree of coverage of the carbon.

10.3 METHODS OF REGENERATION

10.3.1 General Comments

The efficiency of the process of regeneration of the surface of active carbons largely depends on the following factors [3]:

(1) the porous structure of the caron and the chemical condition of its surface,

(2) the physico-chemical properties of the adsorbent,

(3) the methods applied for regeneration,

(4) the conditions under which the regeneration process is conducted.

The regeneration known from the literature can be classified in the following six groups: thermal, chemical (based on extraction and wet oxidation), gas, vacuum, electric and electrochemical and biological.

10.3.2 Thermal and Gas Methods

These methods find widest application in industry and consist of the regeneration of carbon adsorbents with hot gases, primarily superheated steam. These methods are most effective with respect to volatile compounds that neither undergo hydrolysis nor polymerization. Until recently only granular active carbons were subjected to regeneration since they are much more expensive than the powdered carbons used chiefly to decolorize solutions. Granular carbons (crushed or extruded) were returned after use to the manufacturers who regenerated them in the ovens used for the production of active carbon, e.g. internally-fired rotary ovens. However, in view of the significantly increased demand for active carbon used in the treatment of potable water, and also of municipal sewage and industrial waste water, the regeneration of active carbon has become crucial as the amounts of carbon used for these purposes are now very large. Economic reasons have also made it necessary to regenerate powdered active carbons.

Flue gas regeneration is conducted in multiple-hearth furnaces or in rotary ovens at 870–980°C [4, 5]. The temperature of the regenerating agent depends on the properties of the adsorbent–adsorbate system. The residence time of the carbon in the oven is 10 [6] to 30 min [4]. According to Kotzeburo [7], regeneration conducted in rotary ovens requires a 20 per cent excess of air for selective oxidation of organic pollutants with respect to the amount necessary for their stoichiometric combustion. However, if the excess of oxygen is too large, not only the adsorbate but also the carbon is oxidized. The losses of carbon in the regeneration process amount to 4–9% [4].

In the case of regeneration with steam, the process is usually conducted at 100–120°C [8–10] but may also be conducted at higher temperatures [11–14]. The process may be applied to adsorbents in the stationary bed [9–12], moving gravitationally [15], as well as in the fluidized bed [8–14]. If a mixture of gases and steam is used as the regenerating medium, the process is conducted at 750–850°C [15]. If regeneration with steam is impossible, then other hot gases or their mixtures, e.g., nitrogen, carbon dioxide, flue gases or air, are passed through the carbon layer [4, 13, 16].

A diagram of a typical plant for the thermal regeneration of active carbon is shown in Fig. 10.2. It is desirable that the regeneration process be preceded by an initial regeneration of the active carbon in a fluidized bed [4, 12] where control of the temperature and retention time is unnecessary. The effectiveness of this regeneration is influenced by the height of the fluidized bed, the moisture content of the carbon to be regenerated and the composition of the gas mixture used in the process. Figure 10.3 shows the effects on regeneration of variation of temperature and gas composition. As seen from the curves in Fig. 10.3 restoration of the initial adsorptive capacity of the carbon means conducting the process at *ca.* 820°C. The moisture content of the carbon to be regenerated also plays an important role. Steam is used here as a supporting agent. The amounts of steam required differ, however, since in some cases, depending on the type of contaminant on the carbon surface, the use of water may be unnecessary. According to Kotzeburo [7] a 1:1 steam-to-air ratio is quite sufficient.

Sec. 10.3] Methods of regeneration

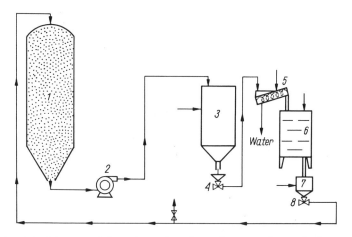

Fig. 10.2 Diagram of the active carbon thermal regeneration cycle: 1 — adsorber, 2 — pump, 3 — oven feeding bin, 4, 8 — ejectors, 5 — draining screen, 6 — oven, 7 — cooling chamber (after Cheremisinoff and Ellerbush [4]).

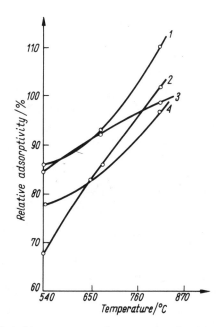

Fig. 10.3 Effect of temperature on the properties of active carbon regenerated in different gaseous media: 1 — $N_2 + H_2O$, 2 — $N_2 + CO_2$, 3 — flue gas, 4 — $N_2 + O_2$ (after Cheremisinoff and Ellerbush [4]).

Table 10.2

Experimental data illustrating losses of carbon in the regeneration process (after Cheremisinoff and Ellerbush [4])

Temperature /°C	Composition of the regenerating gases	Relative adsorption capacity	% loss of carbon
538	air	100	30.8
538	96% N_2:4% O_2	83	3.9
538	98% N_2:2% O_2	78	2.5
649	98% N_2:2% O_2	83	2.4
815	98% N_2:2% O_2	97	2.4

In the course of thermal gaseous regeneration, losses of the regenerated carbon occur. The data summarized in Table 10.2 show that the percentage of carbon lost in the form of carbon dioxide taking place in the regeneration process in air is greatest when the process is conducted at 538°C.

Thermal regeneration cannot be applied if:

(i) Mineral salts are deposited on the carbon surface; in this case preliminary washing is performed with acid, e.g., hydrochloric acid [12].

(ii) The adsorbates present on the carbon surface may pollute the atmosphere after desorption; this refers to such adsorbates as nitrogen oxides, radioactive substances, etc.

(iii) The adsorbates are strongly corrosive after desorption.

Many methods of regeneration of active carbon are combined with its reactivation [17, 18]. Sometimes regeneration is conducted repeatedly over several cycles with recirculation of the regenerating medium [4, 19].

10.3.3 Extraction and Chemical Methods

Thermal regeneration of active carbon necessarily involves the removal of the carbon from the adsorption column, its transport to the regeneration oven, and finally its loading back into the adsorber. These operations are inevitably associated with losses of the carbon which may reach 5%. Among simple, non-thermal regeneration methods, that make it possible to avoid losses of the active carbon during transportation are extraction with special solvents and desorption with the help of chemical reagents.

In the extraction process the adsorbed substances are removed from the carbon with organic solvents. Increase of temperature enhances this process. The substances desorbed in this way may be isolated from the solvent by distillation, extraction, decanting or sedimentation. The carbon is than liberated from the extracting solvent by heating or treatment with steam. An example of such a desorption is the extraction with carbon disulphide of suphur generated on the active carbon during the catalytic oxidation of the adsorbed hydrogen sulphide. The extraction process depends on the substances to be desorbed being readily soluble in the solvent, although some of them may remain in the pores. In contrast, the solvent itself should be completely removable from the pores. The disadvantage of such processes is the need to use large volumes of solvent for complete desorption. Moreover, the solvent then requires purification from the dissolved substances. To reduce substantially the volume of the solution requiring distillation, it is best to use mixtures of dilute solutions of many components as the extracting reagents.

Regeneration with dimethyl ether, which under normal conditions always contains small amounts of water, requires at room temperature a pressure of 0.3–0.5 MPa, and under normal pressure the extraction process must be conducted below $-25°C$. However, the use of dimethyl ether enables regeneration of the active carbon without preliminary drying. The low boiling point of dimethyl ether enables rapid distillation and the presence of water facilitates desorption of hydrophilic materials. It is crucial that the substances to be desorbed be soluble in the liquid ether or water and that in the extraction process no insoluble products are formed on the carbon surface. This kind of extraction is applied to the desorption of phenol, benzene, xylene, alkylbenzenesulphonate-type detergents, many dyes, numerous vitamins, fats and oils.

It is recommended that the materials used as extracting agents for the regeneration of active carbon be polar, protophilic organic solvents with a large dipole moment, e.g. triethanolamine.

The chemical approach to regeneration entails, in general, removing the pollutants from the active carbon surface by virtue of their reaction with suitable chemical reagents and subsequent washing with water in order to remove the regenerating agent. As an example of a commercial application of chemical regeneration, we can cite the removal of phenol from active carbon in the exchange reaction with sodium hydroxide. The Dow Chemical Company (USA) uses in its factory manufacturing phenol, a system of four adsorbers with carbon layers of 13.6 ton weight for purification of waste waters. The individual adsorbers operate periodically. The efficiency of the whole system is 4 to

1150 kg of phenol per minute. Over two years, a 100 adsorption–desorption cycles were completed.

It might seem that losses of carbon are smaller in chemical regeneration than in thermal regeneration. In practice, however, one may find the opposite if the adsorbate and chemical reagents used in regeneration are not fully removed from the carbon surface.

Another example of chemical desorption used on the commercial scale is the removal of acetic acid and *tert*-butylchlorophenol from an active carbon surface after purification of waste waters with sodium hydroxide solution. The use of alkalis and acids is recommended for regeneration in all cases when water-soluble salts are the products of adsorption.

Alkaline regeneration of adsorbents, leading to the removal of acidic adsorbents, is recommended when the active carbon is used for purifying industrial waste waters containing organic acids. Acidic regeneration of adsorbents leads to the desorption of basic substances.

Benzaria and Zundel [9, 10] have described a method of regeneration active carbons used in the purification of waste waters consisting of treating the carbon bed successively with an alkaline solution and an aqueous solution of an organic solvent. After the adsorbed substances are removed, the activated carbon is regenerated with steam and washed with a solution of an inorganic acid.

Milton [20] developed a regeneration method consisting of removing the substances adsorbed on carbon with aqueous or organic solutions of potassium or sodium iodide. This method enables extraction of many organic pollutants.

Yakubeniya *et al.* [21] analysed the effects of active carbon regeneration with various substances, i.e. alcohols, acetone, benzene and alkali metal hydroxide solutions. The carbon dioxide adsorption isotherms show (Fig. 10.4) that the best results are achieved when using boiling aqueous KOH (10%).

In the chemical regeneration of active carbon, inorganic and organic acids such as, e.g., hydrochloric, nitric, acetic, perchloric and bromic acids are used in mixtures with steam [22], as well as hydroxide solutions [3, 9, 10, 21, 23] in aqueous, water-acetone and water-alcohol solvents.

Since hydroxides can remove only certain materials from the active carbon surface, a combined regeneration, consisting of chemical treatment followed by solvent extraction is gaining importance. For instance, the active carbon may be heated directly with steam in an immobile bed

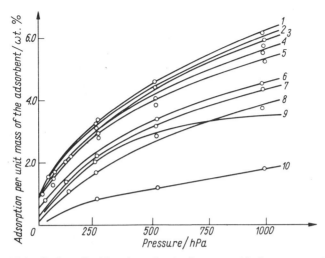

Fig. 10.4 Carbon dioxide adsorption isotherms on Norit regenerated with various solutions: 1 to 5 — 10% potassium hydroxide (regeneration temperature: 1 — 100°C, 2 — 90°C, 3 — 70°C, 4 — 50°C, 5 — 30°C), 6 — alcohol-benzene mixture, 7 — ethanol, 8 — acetone, 9 — methanol, 10 — demineralized carbon (after Yakubeniya et al. [21]).

to 100–110°C and then treated with a 10% sodium carbonate solution. Finally isopropanol is passed slowly through the carbon bed, following which the solvent is removed by passing superheated steam. If necessary, inorganic deposits may be removed by treating the carbon with hydrochloric acid prior to the sodium carbonate treatment. As a result a carbon surface of acidic character is obtained which favours adsorption of organic substances, which are known to be readily adsorbed from aqueous solution at low pH. Pilot plant studies seem to indicate that, in certain cases, chemical regeneration of carbon may prove more effective than thermal treatment.

To show the importance of the regeneration of active carbon, we present the following example. The American Cyanamid Co. has built in its factory in New Jersey, where organic chemicals are manufactured, a plant for waste water treatment on active carbon with a capacity of 75 700 m^3 per day [24]. The plant includes 10 adsorption columns, each of which contains *ca.* 71 tons of active carbon. The multihearth oven used for regeneration returns to the adsorption system only 55 to 69 tons of regenerated active carbon. To rectify the losses about 3.8 tons of fresh active carbon must be added daily. The overal annual running costs of the plant amount to about 5 million dollars.

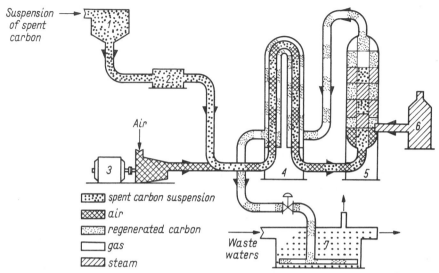

Fig. 10.5 Diagram of the 'wet' oxidation regeneration method: 1 — spent carbon tank, 2 — pump, 3 — air compressor, 4 — heat exchanger, 5 — reactor, 6 — boiler, 7 — adsorption chamber (after Cheremisinoff, Ellerbush [4] and Jankowska et al. [31]).

These various regeneration processes are often conducted at elevated temperatures [25] or under reduced pressure [26]. On the whole, however, chemical methods are less efficient than the thermal methods described earlier.

The so-called wet oxidation is a type of chemical regeneration method. It consists of oxidizing the pollutants adsorbed on the carbon surface, usually with oxygen in the presence of steam (see Fig. 10.5). The process proceeds at 200–300°C for 25–50 min. in autoclaves [4, 14, 27, 28]. In this method the suspension of active carbon to be regenerated is often additionally treated with solutions of inorganic acids. Sometimes the oxidation is conducted in the presence of Cu^{2+} or Cu^{2+} and NH_4^+ cations [29, 30].

10.3.4 Electric and Electrochemical Methods

During regeneration, electric ovens are often used in which the regenerated carbon is a conductor of electric current, which reduces significantly the costs of regeneration. This operation is usually the initial step of a multistep process [32–34] and no destruction of the adsorbate molecules occurs.

Electrochemical methods are also applied to the regeneration of the active carbon surface. As with chemical and extraction methods, the electrochemical methods are very selective with respect to the adsorbates to be removed from the carbon surface. In the electrochemical procedure the spent active carbon is mixed with an electrolyte solution (usually of alkali metal hydroxides) which is then subjected to electrolysis. In this process desorption of the pollutants from the carbon surface occurs, for example, on the anode due to oxidation. The anode or cathode is made of a neutral electrode material [35]. The spent carbon may also be made the cathode when the adsorbate desorbed into the solution is oxidized at the anode [36]. In ref. [3] the carbon regenerated by electrolysis was additionally washed with water, and ca. 80% recovery was achieved.

10.3.5 Other Methods of Regeneration

Vacuum regeneration is often used either in conjunction with other methods or as an individual procedure. However, the vacuum process requires a vacuum-tight apparatus which involves considerable costs. The advantage of the vacuum method is that the adsorbent does not react chemically during the regeneration, nor is diluted with regenerating agents. The pressures involved vary from 1333 to 2666 Pa. In tests on the desorption of dyes at 2666 Pa, 88% efficiency was obtained [26].

Biological regeneration of active carbons is also of practical importance in the purification of industrial waste waters [4, 37, 38]. This process is illustrated schematically in Fig. 10.6. The purification of waste waters consists of the symbiotic action of aerobic and anaerobic bacteria. The initial adsorption of organic pollutants taking place on the active carbon surface is followed by their biological desorption — biodegradation. The organic molecule a undergoes biodegradation to yield

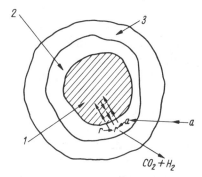

Fig. 10.6 Biological regeneration of active carbon: 1 — particle of active carbon coal, 2 — anaerobic bacterial layer, 3 — aerobic bacterial layer (after Cheremisinoff and Ellerbush [4]).

Table 10.3

Methods of active carbon regeneration according to the type and properties of the substance to be desorbed (after Jankowska et al. [31])

Substance to be desorbed	Method of regeneration	Ref.
Sulphur	600–900°C, neutral gas + hydrogen	[39]
Aniline	nitrogen at 450°C	[40]
o-Nitroaniline	1000°C	[40]
p-Nitrophenol	1000°C	[40]
Sodium dodecyl-benzenesulphonate	liquid ammonia	[41]
Ammonia	condensation and separation	[42, 43]
Sulphur dioxide	MeOH, EtOH aqueous solutions	[44]
Phenol, amines, hydroquinone, etc.	MeOH at 60°C, oxidation at 100–900°C	[45–48]
Sulphur dioxide	firing in a fluidized bed	[49]
Polychlorobiphenyl	refluxing with benzene	[50]

molecule r and this in turn is gradually oxidized biologically to carbon dioxide and water, until simple final products are obtained from carbohydrates, fats, etc.

In Table 10.3 the methods regenerating carbon adsorbents are listed in terms of the type and properties of the substances to be desorbed.

10.4 REGENERATION OF IMPREGNATED CARBON ADSORBENTS

Active carbon featuring surface-deposited additives promoting its chemisorptive and catalytic properties finds wide application as an adsorbent or catalyst. However, it loses this activity over time even when not in use. This loss, due to the action of air and known as ageing, poses an important problem. Several researchers have made attempts to explain the events occurring on the carbon surface [51–54]. Particular interest is focused on active carbon with such surface-activating materials as: chromium, silver, potassium, sodium, cobalt, vanadium and molybdenum salts or organic compounds such as, e.g., pyridine and picoline [55–58]. These sorbents have been used for many years to protect the human respiratory system against highly toxic chemical substances. Particularly important is the chromium-copper sorbent manufactured on a commercial scale. Today this sorbent is discarded when aged, which brings significant losses. The problem exists as to how

Sec. 10.4]   Regeneration of impregnated carbon

to restore its catalytic-sorptive capacity. The previously discussed regeneration methods are unsuitable for these kinds of sorbents in view of their complex catalytic and sorptive system, consisting of the carbon support and the impregnated salt. Little work has been devoted to the regeneration of these sorbents.

Berg and Hjermstad [59] studied the regeneration of a sorbent consisting of approximately 7.0% Cr and 0.05% Ag which was weathered for three years. They applied $\gamma(^{60}Co)$ irradiation in an atmosphere of nitrogen or air under normal reduced pressure with results summarized in Table 10.4. One should note that the dynamic adsorptive capacity of the sorbents was determined with respect to cyanogen chloride of concentration 73 ± 3 mg per 30 dm^3 of dry air at 18°C. Increase of the humidity of the environment in which irradiation was performed hindered regeneration, and the replacement of air by nitrogen slightly improved the regeneration process. The efficiency of regeneration is improved if the pressure is lowered. The authors claim that regeneration by γ-ray irradiation may be applied to sorbents impregnated with various substances, not necessarily metal salts.

Table 10.4

Results of γ-ray regeneration of chromium-copper-silver-impregnated sorbents in dry air (after Berg and Hjermstad [59])

Radiation dose /kJ kg^{-1}	% increase of the filter capacity
0	0
200	1.6
400	7.5
430	32.8
750	61.2
1000	68.2

Some studies on the regeneration of the catalytic and sorptive properties of chromium-copper carbon sorbents have been described in ref. [60]. The concept of the regeneration experiments was based on the results published in [61]. The present authors have determined the dynamic activity and capacity of artificially and naturally weathered chromium-copper sorbents, in parallel with determination of the content of Cr(III) and Cr(VI) in the initial sorbents and in those exposed to weathering. It was found that deterioration of the catalytic and sorptive capacity towards cyanogen chloride due to weathering processes is associated with a decrease in the content of Cr(VI) compounds in favour of Cr(III) via reduction processes proceeding on the carbon surface. This has been confirmed from the quantitative relationships between the duration of the protective activity of the sorbent and the content of chromium compounds on its surface [61]. These conclusions made it possible to propose several regeneration methods based on a reversal of the reduction of Cr(VI) to Cr(III). Account has also been taken of the conclusions drawn in ref. [23] that a basic medium favours the regeneration of these sorbents. An additional crucial problem is avoidance of any changes in the porous structure of the sorbent during regeneration which would reduce its adsorptive properties.

One regeneration method [60] consisted of impregnating the sorbents with hydrogen peroxide and ammonia solutions and then heating them at 150°C for 3 h. The best effects were obtained when a solution of 1:2 hydrogen peroxide to aqueous ammonia ratio was used.

The possibilities of regeneration with potassium permanganate solutions [62] and supersonic waves [63] have also been studied. Good results have been obtained when hydrogen peroxide and ammonia solutions were used. It was found that the regenerated sorbent had properties as good as fresh material (if not better).

The laboratory tests were confirmed in industry where an attempt was undertaken to regenerate a naturally weathered chromium-copper sorbent using hydrogen peroxide and aqueous ammonia in a 1:2 ratio. The tests were carried out in a plant used for the production of sorbents. Investigations of the catalytic and sorptive properties of the regenerated sorbent have shown [63] that the latter meets commercial requirements and may be used again successfully. Analysis of the costs of regeneration has shown that these amount to about 50% of the manufacturing costs of a new sorbent, which is promising.

References

[1] Klein, J., *Staub-Reinhalt-Luft* **36**, 292 (1976).
[2] Kienle (von), H., Bäder, E., *Activkohlen und ihre industrielle Anwendungen*, Ferdinand Enke Verlag, Stuttgart 1980.
[3] Kusajko, W., Mortka, S., *Biul. WAT* No. 4, **29**, 81 (1980).
[4] Cheremisinoff, P. N., Ellerbush, F., *Carbon Adsorption Handbook*, Ann Arbor Science Publishers, Ann Arbor, Mich. 1978.
[5] Zanitsh, R. H., Lynch, R. T., *Chem. Eng.* **85**, 95 (1978).
[6] Kulev, A. M., Agakishev, N. A., Pinsker, B. A., *Gaz. Prom.* **4**, 48 (1964).
[7] Kotzeburo, N., *Kiorocu Juki Kogo Kenk* **51**, 332 (1976).

[8] Antsypovich, J. S., Shkatov, J. F., *Khim. Prom.* **3**, 57 (1973).
[9] French Patent 2 094 336, 1972.
[10] GDR Patent 83 755, 1971.
[11] Japanese Patent 79 147 173, 1979.
[12] Kunitara, K., Toshio, O., *Seisan Kenkyu* **28**, 109 (1976).
[13] Teroda, K., Kudo, K., Mitooko, H., et al. *Sekiyu Gakkai Shi* **21**, 221 (1978).
[14] French Patent 2 261 976, 1975.
[15] USSR Patent 343 950, 1972.
[16] Japanese Patent 78 113 287, 1978.
[17] Matsumoto, Z., Numasaki, *Hydrocarbon Process* **55**, 157 (1976).
[18] Japanese Patent 7 895 195, 1978.
[19] British Patent 1 300 744, 1972.
[20] USA Patent 4 058 457, 1977.
[21] Yakubeniya, N. A., Lukin, V. D., Astakhov, V. A., et al., *Zh. Prikl. Khim.* **52**, 2539 (1979).
[22] USA Patent 4 038 301, 1977.
[23] Beccari, M., Paolini, A. E., Variali, G., *Efflum. a. Wat. Treatm. J.* **17**, 287 (1977).
[24] *Chemical Week*, September, 31 (1976).
[25] Japanese Patent 76 131 496, 1976.
[26] Japanese Patent 76 126 393, 1976.
[27] Gitchel, W. G., Meidl, J. A., Buranth, W., *Chem. Eng. Progr.* **71**, 90 (1975).
[28] Jahning, C., GwF-wasser-abwasser **74**, 28 (1978).
[29] Japanese Patent 75 102 587, 1975.
[30] Japanese Patent 7 556 394, 1975.
[31] Jankowska, H., Choma, J., Kozaczyński, W., *Przem. Chem.* **62**, 69 (1983).
[32] Kanegawa, M., *Ebara Infuriko Jiho* **74**, 28 (1978).
[33] Japanese Patent 7 916 475, 1979.
[34] Australian Patent 420 559, 1972.
[35] South Africa Patent 7 704 287, 1978.
[36] USSR Patent 505 610, 1976.
[37] Balice, V., Baori, G., Liberti, L., *Inquinamento* **19**, 61 (1977).
[38] USA Patent 3 803 029, 1974.
[39] Japanese Patent 7 649 196, 1976.
[40] Amicarelli, V., Baldassare, G., Liberti, L., *J. Therm. Anal.* **18**, 155 (1980).
[41] Japanese Patent 7 689 888, 1976.
[42] Japanese Patent 7 743 795, 1977.
[43] Japanese Patent 7 691 893, 1976.
[44] Japanese Patent 7 789 592, 1977.
[45] Japanese Patent 76 139 593, 1976.
[46] Japanese Patent 7 737 597, 1977.
[47] Japanese Patent 7 651 151, 1976.
[48] Japanese Patent 7 654 489, 1976.
[49] FRG Patent 2 460 312, 1976.
[50] Japanese Patent 74 122 895, 1974.
[51] Tomassi, W., Neffe, S., *Biul. WAT* **22**, No. 12, 256 (1973).
[52] Swinarski, A., Żytkowicz, A., *Chem. Stosowana* **19**, 421 (1975).
[53] Choma, J., Doctoral Thesis, Military Technical Academy, Warsaw 1981.
[54] Hjermstad, H. P., Berg, R., *Am. Ind. Hyg. Assoc. J.* **38**, 211 (1977).

[55] USA Patent 1 789 194, 1930.
[56] French Patent 1 575 501, 1969.
[57] Poziomek, E. J., Mackay, R. A., Barett, R. P.., *Carbon* **13**, 259 (1975).
[58] Smišek, M., Černý, S., *Active Carbon*, Elsevier, Amsterdam; London; New York 1970.
[59] FRG Patent 2 335 297, 197.
[60] Choma, J., Jankowska, H., Kozaczyński, W., et al. *Biul. WAT* **31**, No. 3, 97 (1982).
[61] Jankowska, H., Choma, J., Kozaczyński, W., *Przem. Chem.* **60**, 320 (1981).
[62] Jankowska, H., Choma, J., Kozaczyński, W., *ibid.* **62**, 518 (1983).
[63] Kozaczyński, W., Doctoral Thesis, Military Technical Academy, Warsaw 1983.

Index

absolute density, 53
activation
  chemical, 13, 52
  general, 13, 38
  steam gas, 13
activation of brown coal, 24
active carbon
  electrode material, 250
  electrode properties, 100
  medicinal uses, 246
  nature, 9
  structure, 75
additives, 80
adsorption
  energetics, 163
  from gas phase, 219
  from liquid phase, 170, 231
  heat of, 163
  isotherm (of water vapour), 97
  manostat, 200
  measurements, 200
  methods, 193
  models for, 107
  of electrolytes, 100
  of polar adsorbates from gas phase, 96
  on active carbons, 180
  potential theory of, 128
  tests, 56
adsorption liquid microburettes, 196
ash, 80
ash content, 60

BET isotherm, 113
binary liquid solutions, 175
binding materials, 28
black peat, 24
blood purification, 248
brown coal activation, 214
bulk density, 53

calorimeters, 167
capillary condensation, 121
carbon reaction with water vapour, 40, 41
carbonization, 31
  granules, 36
  temperature, 33
carbons
  commercial, 16
  laboratory, 17
  oxidation, 21
charcoal, 25
chemical testing, 60
coals, classification, 20
commercial active carbon, 16

decolorizing oils and fats, 235
Deguss process, 25
density, 53
D–R equation, 133, 143

electrode properties, 100

Index

electron microscopy, 217
extruded carbons, 54

fluidized bed method, 37
fluidized bed ovens, 44
Freundlich isotherm, 112

gamma distribution function, 198
Gibbs adsorption isotherm, 172
granulation, 29
graphite, 76
$G(x)$ function, 152

Halsey equation, 124
hard coal, 19
hardness, 56
Harkins–Jura equation, 120
Harkins–Jura relative method, 120
helium method, 210
Henry's law, 108
heteroatoms in active carbon, 81

iodine adsorption, 57
ion-exchangers, 256

Jaroniec–Choma equation, 149

laboratory carbons, 17
Langmuir isotherm, 110
lignite, 24

macropore, 132
mercury porosimetry, 212
mesopore, 132
   surface area, 136
methyl blue adsorption, 58
micropore, 132
   surface area, 153
   thermodynamics of adsorption at, 155
micropore filling theory, 141, 144
microporosity estimation, 133
microporous structure
   heterogeneity, 145
   filling of, 146

milligram value, 60
mineral additives, 80
moisture content, 60

Norit carbons, 61

particle density, 53
peat coke, 24
phenazone adsorption, 59
phenol adsorption isotherm, 56
pH value, 61
physical testing, 53
pore volume, 207
porosimetry, mercury, 212
porous structure, complete analysis of, 158
powdered carbons, 54
production, general, 13

raw materials, 21
regeneration, 260
   extraction and chemical methods, 268
   thermal and gas methods, 265
rotary oven, 45

sampling, 52
Sific process, 25
SO_2 removal, 220
sorption spiral balances, 194
steam activation, 41
surface
   chemical nature, 81
   cyclic voltammetry, 92
   EPR studies, 96
   functionality, 82
   functional group analysis, 85
   polarographic studies, 92

t-method, 132

upper respiratory tract, protection of, 225

waste treatment, 237
waste water treatment, 237